# Light Pulse Compression

# Laser Science and Technology
## An International Handbook

## Editors in Chief

V.S.LETOKHOV, *Institute of Spectroscopy, USSR Academy of Sciences, 142092 Moscow Region, Troitsk, USSR*

C.V. SHANK, *AT&T Bell Laboratories, Holmdel, NJ 07733, USA*

Y.R.SHEN, *Department of Physics, University of California, Berkeley, CA 94720, USA*

H. WALTER, *Max-Planck-Institute für Quantenoptik und Sektion Physik, Universität München, D-8046 Garching, FRG*

**Volumes in preparation are listed on the inside back cover**

This book is part of a series. The publisher will accept continuation orders which may be cancelled at any time and which provide for automatic billing and shipping of each title in the series upon publication. Please write for details.

# Light Pulse Compression

Wolfgang Rudolph
Bernd Wilhelmi
*Friedrich Schiller University Jena*
*German Democratic Republic*

harwood academic publishers
chur.london.paris.new york.melbourne

PHYSICS

© 1989 by Harwood Academic Publishers GmbH, Poststrasse 22, 7000 Chur, Switzerland. All rights reserved.

Harwood Academic Publishers

| | |
|---|---|
| Post Office Box 197 | Post Office Box 786 |
| London WC2E 9PX | Cooper Station |
| England | New York, New York 10276 |
| | United States of America |
| | |
| 58, rue Lhomond | Private Bag 8 |
| 75005 Paris | Camberwell, Victoria 3124 |
| France | Australia |

**Library of Congress Cataloging-in-Publication Data**

Rudolph, Wolfgang, 1956-
    Light pulse compression/Wolfgang Rudolph, Bernd Wilhelmi.
      p.    cm. — (Laser science and technology)
    Includes index.
    ISBN 3-7186-4888-1
    1. Laser pulses, Ultrashort.  2. Laser spectroscopy.  3. Nonlinear optics.  I. Whilhelmi, Bernd.  II. Title.  III. Series.
QC689.5.L37R83   1989
621.36'6—dc19   '                        89–1772
                                                        CIP

# Contents

# Introduction to the Series

Almost 30 years have passed since the laser was invented; nevertheless, the fields of lasers and laser applications are far from being exhausted. On the contrary, during the last few years they have been developing faster than ever. In particular, various laser systems have reached a state of maturity such that more and more applications are seen suffusing fields of science and technology, ranging from fundamental physics to materials processing and medicine. The rapid development and large variety of these applications call for quick and concise information on the latest achievements; this is especially important for the rapidly growing interdisciplinary areas.

The aim of "Laser Science and Technology — An International Handbook" is to provide information quickly on current as well as promising developments in lasers. It consists of a series of self-contained tracts and handbooks pertinent to laser science and technology. Each tract starts with a basic introduction and goes as far as the most advanced results. Each should be useful to researchers looking for concise information about a particular endeavor, to engineers who would like to understand the basic facts of the laser applications in their respective occupations, and finally to graduate students seeking an introduction into the field they are preparing to engage in.

When a sufficient number of tracts devoted to a specific field have been published, authors will update and cross-reference their pages for publication as a volume of the handbook.

All the authors and section editors are outstanding scientists who have done pioneering work in their particular field.

*V.S. Letokhov*
*C.V. Shank*
*Y.R. Shen*
*H. Walther*

# LIGHT PULSE COMPRESSION

W. RUDOLPH and B. WILHELMI

*Friedrich-Schiller-University Jena, Physics Department, JENA*
*German Democratic Republic*

## 1. INTRODUCTION

Considerable progress has taken place in the last two decades in the generation of ultrashort light pulses (see e.g., [1.1–2.2] and Fig. 1.1). Until 1965 the duration of the shortest light pulses remained at about 1 nanosecond and was determined by the most advanced electronic techniques of that time.

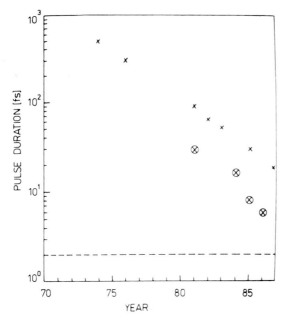

**Figure 1.1** Development of the shortest achieved pulse durations direct from the laser ( × ) and after external compression (⊗) as a function of the year (from [1.1]).

Ultrashort light pulses were first generated in 1965 by passive mode-locking of a ruby laser [1.3] and one year later the same method was successfully applied to Nd:glass lasers [1.4], where pulse durations of only some picoseconds were achieved. From that time on, the minimum duration of light pulses has no longer been determined by electronic methods and techniques, but on the contrary, the shortest electronic pulses are nowadays generated and measured by the use of light pulses [1.5]. In the meantime, several methods have been developed in order to generate ever shorter light pulses at wavelengths from the ultraviolet to the infrared.

In 1981, the first light pulse with a duration of less than 0.1 picoseconds or 100 femtoseconds were generated by improvements of the passively mode-locked dye laser [1.6] using colliding pulse laser configurations. At present the shortest light pulses directly emitted from lasers have durations of about 30 fs. They are obtained by utilizing additional dispersive linear optical elements in such dye lasers for intracavity pulse compression (see [1.7–1.12] and Section 5).

The shortest light pulses yet reported have a duration of 6–8 fs, i.e., they consist of only three to four wave cycles of their center wavelength of 0.6 $\mu$m [1.13]. They were obtained by amplifying the light pulses from a passively mode-locked ring dye laser and by passing the amplified pulses through a nonlinear optical medium, which produces phase modulation connected with spectral broadening. This phase modulation consists of an increase of the instantaneous frequency, denoted as frequency sweep or chirp, which proceeds almost linearly. It is then compensated by passing the light pulses through an appropriate loss-free dispersive linear optical device that per definition does not change the spectral content of the transmitted light. Hence, at the end, compressed light pulses are achieved, the duration of which is on the order of the inverse value of the bandwidth that had been increased by the nonlinear optical means before (see Section 4).

From these examples we see that the shortest laser pulses as well as the shortest light pulses of all have been produced by compression methods (cf. Fig. 1.1). Pulse compression is, however, not only a necessary tool for breaking records, but, moreover, it is at present a very practical, not too expensive and widely used experimental technique to produce short light pulses at a variety of wavelengths and for different applications. For these reasons pulse compressors became also commercially available in the last years [1.14]. The compression of light pulses after chirp production in nonlinear optical media was already proposed in 1969 [1.15–1.18] and since that time it has been improved in many laboratories (see e.g., [1.19–1.30], [1.13]). Until 1980 the frequency sweep had been produced in bulk media where it is difficult to achieve a large linear increase of the instantaneous frequency. The application of optical fibers was only discussed in connection with the soliton-like pulse propagation as a way to overcome the pulse spreading due to the dispersion i.e., for the wavelength range where group velocity dispersion and nonlinearity have opposite signs (e.g., [1.31] and references therein). A very important improvement was the introduction of single-mode fibers as nonlinear optical media [1.21] for the other wavelength range, where with rather large interaction lengths the interplay of dispersion and nonlinearity produces an almost linear and large frequency chirp.

The rapid development of sources for picosecond and femtosecond light pulses and in particular of pulse compression stimulated an enormous progress in the entire field of ultrafast measuring technology (see e.g., [1.1], [1.2], [1.29–1.35]). Ultrashort light pulses permit novel spectroscopic investigations of extremely rapidly proceeding physical, chemical and biological phenomena in the ultraviolet, the visible and the infrared spectral regions. Among the fundamental processes that have been measured on the femtosecond time scale are the decay of molecular vibrations and phonons, electronic relaxation processes in molecules and solids, energy migrations, charge transfer, isomerization, reorientation of molecules. Moreover, picosecond and femtosecond technology provides possibilities for the manipulation of photophysical and photochemical processes, phase transitions included. Starting from intense ultrashort light pulses, it is furthermore possible to generate X-ray pulses [1.36], and electron pulses of comparable duration, the latter are produced in vacuo [1.37], [1.38] as well as in semiconductors [1.5]. By use of such techniques, electronic and optoelectronic devices can be rapidly modulated, and their temporal response can be measured with high resolution. The aims of the specific application determine the desired values of the light pulse parameters to be applied, e.g., of wavelength, pulse duration, pulse shape and frequency modulation. Increasingly chirp production and chirp compensation along with pulse shaping and in particular with pulse compression within and outside the laser cavity are applied to meet the various requirements.

The aim of this paper is to review essential experimental results and theoretical concepts of intracavity and extracavity light pulse compression based on chirp generation and compensation. The structure of the paper shall be as follows. We begin with the characterization of ultrashort light pulses and explain basic principles of pulse propagation and pulse compression (chapter 2). This is followed by the description of various linear optical elements and their influence on light pulses (chapter 3). Chapter 4 deals with the detoriation of amplitude and phase in various dispersive, nonlinear samples and their importance for extracavity pulse compression. Then intracavity pulse compression is discussed for passive mode-locked dye lasers and soliton lasers (chapter 5). Finally certain aspects of the measurement of and with chirped pulses are considered (chapter 6).

## 2. DESCRIPTION OF THE ULTRASHORT LIGHT PULSES, DEFINITIONS AND NOTATIONS

### 2.1 Fourier transforms

Light pulses may be described completely in either the time domain or the frequency domain, where, e.g., the electric field strength is denoted as E(t) and $\underline{E}(\omega)$, respectively. (More thoroughly speaking, E denotes one of the vector components of the field strength, where we have omitted the vector index for the sake of simplicity of the equations.) Both are connected by

$$E(t) = \frac{1}{2\pi} \int_{-\infty}^{\infty} d\omega \, \underline{E}(\omega)e^{i\omega t} \tag{2.1a}$$

and

$$\underline{E}(\omega) = \int_{-\infty}^{\infty} dt E(t)e^{-i\omega t} \tag{2.1b}$$

The complex Fourier transform $\underline{E}(\omega)$ may be expressed as

$$\underline{E}(\omega) = a(\omega)e^{i\phi(\omega)} \tag{2.2}$$

where $a(\omega)$ and $\phi(\omega)$ are, respectively, called the spectral amplitude and phase.

Sometimes it is useful to separate the negative frequency part of the field strength from the positive one:

$$E(t) = E^{(-)}(t) + E^{(+)}(t) \tag{2.3a}$$

where

$$E^{(-)}(t) = \frac{1}{2\pi} \int_{-\infty}^{0} d\omega \underline{E}(\omega)e^{i\omega t} = \frac{1}{2\pi} \int_{0}^{\infty} d\omega \underline{E}(-\omega)e^{-i\omega t} \tag{2.3b}$$

$$E^{(+)}(t) = \frac{1}{2\pi} \int_0^\infty d\omega \underset{\sim}{E}(\omega) e^{i\omega t} \tag{2.3c}$$

$$E^{(-)}(t) = [E^{(+)}(t)]^* \tag{2.3d}$$

and

$$\underset{\sim}{E}{}^{(+)}(\omega) = \int_{-\infty}^\infty dt E^{(+)}(t) e^{-i\omega t} = \begin{cases} \underset{\sim}{E}(\omega) & \text{for } \omega \geqslant 0 \\ 0 & \text{for } \omega < 0 \end{cases} \tag{2.3e}$$

In our notation $E^{(+)}(t)$ may be used as the complex analytic signal (see e.g., [2.1]).

## 2.2 Carrier frequency and pulse envelope

If the bandwidth of the light pulse is small compared to its center frequency, it is useful to introduce the concept of carrier frequency and pulse envelope. The electric field is then represented as

$$E(t) = \frac{1}{2} \overline{E}(t) e^{i\omega_L t} + \text{c.c.} \tag{2.4a}$$

where

$$\overline{E}(t) = A(t) e^{i\Phi(t)} \tag{2.4b}$$

is the complex pulse envelope (or the complex instantaneous pulse amplitude); $A(t)$ and $\Phi(t)$ are the instantaneous amplitude (modulus) and phase of the light pulse. The instantaneous frequency $\omega(t)$ is defined as

$$\omega(t) = \omega_L + \delta\omega(t) \tag{2.4c}$$

where

$$\delta\omega(t) = \frac{d}{dt} \Phi(t) \tag{2.4d}$$

Note that the decomposition of $\omega(t)$ into $\omega_L$ and $\delta\omega(t)$ is not unique. In what follows, we identify $\omega_L$ mostly with the instantaneous frequency at the pulse maximum and call it mid or center frequency. The pulse is phase modulated if

$$\Phi(t) \neq \text{const.} \tag{2.4e}$$

and frequency modulated or chirped if

$$\frac{d}{dt} \Phi(t) \neq const. \tag{2.4f}$$

We will call the pulse down (up) chirped if $\frac{d^2}{dt^2} \Phi(t) \overset{<}{(>)} 0$ in the vicinity of the pulse maximum. Sometimes it is convenient to use averaged values of the phase modulation and chirp which are defined as

$$\overline{\frac{d^n}{dt^n} \Phi(t)} = \frac{\displaystyle\int_{-\infty}^{\infty} |\overline{E}(t)|^2 \frac{d^n\Phi(t)}{dt^n} dt}{\displaystyle\int_{-\infty}^{\infty} |\overline{E}(t)|^2 dt} \tag{2.4g}$$

The instantaneous light intensity I(t), i.e., the power per area, and the photon flux $J(t) = I(t)/\hbar \omega_L$ are given by

$$I(t) = 2\epsilon_0 c E^{(-)}(t) E^{(+)}(t) = \tfrac{1}{2} \epsilon_0 c A^2(t) = \tfrac{1}{2} \epsilon_0 c |\overline{E}(t)|^2 \tag{2.5}$$

The pulse duration $\tau_L$ is here defined as the full width at half maximum (FWHM) of I(t) and J(t).

The spectral intensity, which is measured by use of spectrometers without time resolution is given by

$$\underline{I}(\omega) = \frac{\epsilon_0 c}{\pi} |\underline{E}(\omega)|^2 = \frac{\epsilon_0 c}{\pi} a^2(\omega) \tag{2.6}$$

The spectral bandwidth (FWHM) of the pulse is denoted by $\Delta\omega$. The possible values of the pulse duration bandwidth product

$$p = \frac{\Delta\omega}{2\pi} \tau_L \tag{2.7a}$$

are restricted by

$$p \geqslant c_B \tag{2.7b}$$

where $c_B$ is a figure of the order of unity, the value of which is characteristic for the temporal profile of the particular light pulse (see Table 2.1). With $p = c_B$ the pulse shows no frequency modulation at all ($\dot{\Phi}(t) = const$) and is called transform-limited or bandwidth-limited; this pulse is the shortest one possible for a given spectral bandwidth and shape. It should also be noted that a bandwidth limited pulse always has a symmetrical amplitude spectrum as it can easily be shown.

**Table 2.1** Pulse profile parameter $c_B$

| Name | Intensity profile | Value of $c_B$ |
|------|-------------------|----------------|
| Gaussian | $\exp(-4 \ln 2(t/\tau_L)^2)$ | 0.441 |
| Lorentzian | $[1 + (2t/\tau_L)^2]^{-1}$ | 0.221 |
| sech$^2$ | $\cosh^{-2}(1.76t/\tau_L)$ | 0.315 |

As an example we give the field strength of a Gaussian light pulse with linear frequency chirp:

$$E(t) = \frac{1}{2} A(t)e^{i\Phi(t)}e^{i\omega_L t} + \text{c.c.} \qquad (2.8a)$$

with

$$A(t) = A_0\, e^{-\gamma_0 t^2} \qquad (2.8b)$$

and

$$\Phi(t) = \Phi_0 + \beta_0 t^2 \qquad (2.8c)$$

where the pulse duration is given by

$$\tau_{L0} = \sqrt{\frac{2 \ln 2}{\gamma_0}} \qquad (2.8d)$$

The Fourier transform is given by

$$\underline{E}_0^{(+)}(\omega) = \frac{\sqrt{\pi}}{2} A_0 \frac{1}{\sqrt{\gamma_0 - i\beta_0}}\, e^{i\Phi_0} \exp\left[ -\frac{(\omega - \omega_L)^2}{4(\gamma_0 - i\beta_0)} \right] \qquad (2.8e)$$

i.e.,

$$a(\omega) = \frac{\sqrt{\pi}}{2} A_0 \frac{1}{\sqrt[4]{\gamma_0^2 + \beta_0^2}} \exp\left[ -\frac{\gamma_0(\omega - \omega_L)^2}{4(\gamma_0^2 + \beta_0^2)} \right] \qquad (2.8f)$$

and

$$\phi(\omega) = \Phi_0 + \frac{1}{2} \arctan \frac{\beta_0}{\gamma_0} - \frac{\beta_0(\omega - \omega_L)^2}{4(\gamma_0^2 + \beta_0^2)} \qquad (2.8g)$$

We call $\beta_0$ the chirp parameter, which is related to $\delta\omega(t)$ of equation (2.4d) by

$$\delta\omega(t) = \frac{d}{dt} \Phi(t) = 2\beta_0 t \qquad (2.8h)$$

The spectral bandwidth is

$$\Delta\omega = \sqrt{8 \ln 2 \left(\frac{\gamma_0^2 + \beta_0^2}{\gamma_0}\right)} \tag{2.8i}$$

and

$$p = \frac{2 \ln 2}{\pi} \left[1 + \left(\frac{\beta_0}{\gamma_0}\right)^2\right]^{1/2} \geq \frac{2 \ln 2}{\pi} \approx 0.441 = c_B \tag{2.8j}$$

(see also Figure 2.1).

## 2.3 Spatial Fourier transform and wave envelopes

Analogous to the time dependence, we describe the spatial dependence of the field either by

$$E(t, \vec{r}) = \left(\frac{1}{2\pi}\right)^4 \int_{-\infty}^{\infty} d\omega \int_{-\infty}^{\infty} d^3\vec{k} \, \underset{\sim}{E}(\omega, \vec{k}) e^{i(\omega t - \vec{k}\vec{r})} \tag{2.9a}$$

or by its Fourier transform

$$\underset{\sim}{E}(\omega, \vec{k}) = \int_{-\infty}^{\infty} dt \int_{-\infty}^{\infty} d^3\vec{r} \, E(t,\vec{r}) e^{-i(\omega t - \vec{k}\vec{r})} \tag{2.9b}$$

If not only the bandwidth $\Delta\omega$ is small compared to the carrier frequency $\omega_L$, but also the modulus of the wave vector interval $\Delta k = |\Delta\vec{k}|$ compared to $k_L = |\vec{k}(\omega_L)|$, it is useful to introduce the wave envelope $\bar{E}(t, \vec{r})$ by

$$E(t, \vec{r}) = \tfrac{1}{2} \bar{E}(t, \vec{r}) \, e^{i(\omega_L t - \vec{k}_L \vec{r})} + \text{c.c.} \tag{2.10}$$

where the Fourier transform of the envelope $\underset{\sim}{\bar{E}}(\Omega, K) = \mathscr{L}\{\bar{E}(t, \vec{r})\}$ is related to $\underset{\sim}{E}^{(+)}(\omega, \vec{k})$ by

$$\underset{\sim}{E}^{(+)}(\omega, \vec{k}) = \tfrac{1}{2} \underset{\sim}{\bar{E}}(\omega - \omega_L, \vec{k} - \vec{k}_L) \tag{2.11}$$

## 2.4 Wave equation

The pulse propagation through a medium obeys the wave equation (see e.g. [2.2]).

$$\left(-\frac{\partial^2}{\partial z^2} + \frac{1}{c^2} \frac{\partial^2}{\partial t^2}\right) E(t,z) = -\mu_0 \frac{\partial^2}{\partial t^2} P(t,z) \tag{2.12}$$

For simplification, we assumed here a linearly polarized plane wave travel-ing in the z-direction. The polarization $P(t,z)$ describes the interaction

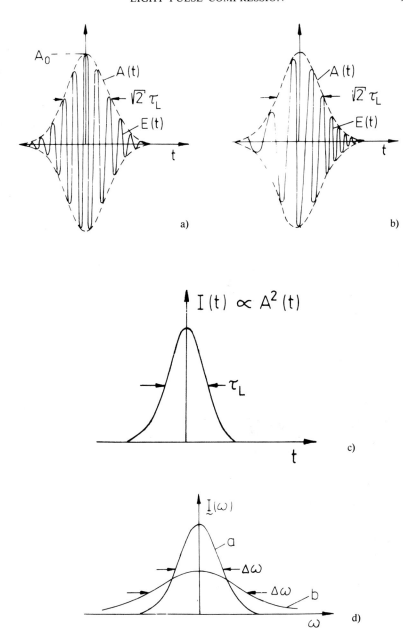

**Figure 2.1** Gaussian light pulse with and without chirp, a. electric field strength of a bandwidth-limited pulse with Gaussian profile of duration $\tau_L$, b. electric field strength of a chirped pulse with Gaussian profile of duration $\tau_L$, c. intensity of the pulses from a) and b), d. spectral intensity of the pulses from a) and b).

with the medium and is usually decomposed into a linear (L) and a non-linear (NL) term (see e.g. [2.3], [2.4])

$$P(t,z) = P^L(t,z) + P^{NL}(t,z) \qquad (2.13)$$

In general, (2.12) is a nonlinear partial (integro-) differential equation of the second order which can hardly be solved without approximations even by means of numerical methods. Therefore, it is useful to simplify (2.12) by utilizing the characteristics of the pulse propagation in the actual medium. In the following we will derive a reduced wave equation for the pulse envelope (2.10), where we account only for a linear polarization. This describes the pulse propagation through a linear optical medium which often can be considered as the host medium for sources of a non-linear polarization. The effect of $P^{NL} \neq 0$ will then be discussed in Section 4 in detail.

The linear part of the polarization $P^L(t,z)$ is related to the electric field by the integral

$$P^L(t,z) = \epsilon_0 \int_{-\infty}^{\infty} dt' \; \kappa(t') \; E(t-t',z) \qquad (2.14a)$$

where $\kappa(t')$ is the response function of the material. Here we assumed that spatial dispersion of the sample can be neglected. The Fourier transform of (2.14a) reads

$$\underline{P}^L(\omega,z) = \epsilon_0 \underline{\kappa}(\omega) \, \underline{E}(\omega,z) \qquad (2.14b)$$

$$\underline{P}^L(\omega,k) = \epsilon_0 \underline{\kappa}(\omega) \, \underline{E}(\omega,k) \qquad (2.14c)$$

where

$$\underline{\kappa}(\omega) = \epsilon(\omega) - 1 \qquad (2.14d)$$

is the susceptibility of the medium and $\epsilon(\omega)$ is its dielectric permittivity.

Sometimes it is useful to solve the linear wave equation in the frequency domain, in particular if $\underline{\kappa}(\omega)$ changes rapidly within frequency ranges comparable with the pulse bandwidth. (As we will see later, in the time domain this would require the consideration of additional time derivatives). In the frequency domain the wave equation can be written with (2.14b,d) as

$$\left( -\frac{\partial^2}{\partial z^2} - \frac{\omega^2}{c^2} \epsilon(\omega) \right) \underline{E}(\omega,z) = 0 \qquad (2.15a)$$

which has the general solution

$$\underline{E}(\omega,z) = \underline{E}(\omega,z=0)e^{-ik(\omega)z} \qquad (2.15b)$$

where $k(\omega)$ satisfies the dispersion relation of linear optics

$$k^2(\omega) = \frac{\epsilon(\omega)}{c^2} \, \omega^2 \tag{2.15c}$$

It should be noted that $\epsilon(\omega)$ can be complex, resulting in a complex propagation constant $k(\omega) = k_r(\omega) + i \, k_i(\omega)$. In this notation $k_i$ is responsible for linear losses and $k_r$ describes the effect of lossless dispersion.

If the bandwidth of the pulse is small compared to its center frequency, in analogy to (2.10), it is convenient to represent the polarization by an envelope and a rapidly oscillating term

$$P^L(t,z) = \tfrac{1}{2} \, \overline{P}^L(t,z) e^{i(\omega_L t - k_L z)} + \text{c.c.} \tag{2.16a}$$

and we may expand $\underline{P}^L(\omega,z)$ of (2.14b) at $\omega = \omega_L$. Back transform to the time domain yields for the envelope

$$
\overline{P}^L(t,z) = \epsilon_0 \left\{ \underline{\kappa}_L - i \, \underline{\kappa}'_L \frac{\partial}{\partial t} - \frac{1}{2} \, \underline{\kappa}''_L \frac{\partial^2}{\partial t^2} - \frac{i}{6} \, \underline{\kappa}'''_L \frac{\partial^3}{\partial t^3} \right.
$$
$$
\left. + \frac{1}{24} \, \underline{\kappa}''''_L \frac{\partial^4}{\partial t^4} + \ldots \right\} \overline{E}(t,z) \tag{2.16b}
$$

where

$$\underline{\kappa}_L = \underline{\kappa}_L(\omega_L), \; \underline{\kappa}'_L = \left. \frac{d\underline{\kappa}}{d\omega} \right|_{\omega = \omega_L}, \; \text{etc.} \tag{2.16c}$$

The use of (2.16) in (2.12) gives

$$
\left( 2i \, k_L \frac{\partial}{\partial z} - \frac{\partial^2}{\partial z^2} \right) \overline{E}(t,z)
$$
$$
+ \frac{2i \, \omega_L}{c^2} \left( 1 + \underline{\kappa}_L + \frac{1}{2} \, \omega_L \underline{\kappa}'_L \right) \frac{\partial}{\partial t} \, \overline{E}(t,z)
$$
$$
+ \frac{1}{c^2} \left( 1 + \underline{\kappa}_L + 2\omega_L \underline{\kappa}'_L + \frac{1}{2} \, \omega_L^2 \underline{\kappa}''_L \right) \frac{\partial^2}{\partial t^2} \, \overline{E}(t,z)
$$
$$
- \frac{i}{c^2} \left( \underline{\kappa}'_L + \omega_L \underline{\kappa}''_L + \frac{1}{6} \, \omega_L^2 \underline{\kappa}'''_L \right) \frac{\partial^3}{\partial t^3} \, \overline{E}(t,z) + \ldots = 0 \tag{2.17}
$$

Note that the second derivative of the propagation constant $k$ is related to $\underline{\kappa}_L$ by

$$k''_L = \left. \frac{d^2 k}{d\omega^2} \right|_{\omega = \omega_L} = \frac{1}{2c\sqrt{\underline{\kappa}_L + 1}} \left[ 2\underline{\kappa}'_L + \omega_L \underline{\kappa}''_L - \frac{\omega_L (\underline{\kappa}'_L)^2}{2(1 + \underline{\kappa}_L)} \right] \tag{2.18}$$

To simplify the wave equation (2.17) we assume

$$\Delta\omega/\omega_L \ll 1 \qquad (2.19a)$$

and

$$\Delta k/k_L \ll 1 \qquad (2.19b)$$

which justifies to neglect spatial and temporal derivatives of $\overline{E}(t,z)$ compared to $|k_L \overline{E}(t,z)|$ and $|\omega_L \overline{E}(t,z)|$, respectively. This approximation is often called slowly varying envelope approximation (SVEA). In particular the term with the second spatial derivative is of minor influence compared to others. (For an estimation of the order of magnitude of this term we use the solution of the wave equation (2.15) for Gaussian pulses (see section 3.1) traversing a material where $\varkappa_L^{(n)}$ can be neglected for $n > 2$. It can easily be shown that the term with the second spatial derivative is on the order of $(10k_L''/\tau_L^2)^2$, and the term with the second spatial derivative divided by the term with the first spatial derivative is on the order of $10(\omega_L\tau_L^2)/(k_L''c)$. For unchirped 10fs pulses at 0.6 $\mu$m we have e.g. $\omega_L\tau_L = 30$ and in glass with $k_L'' \simeq 6 \ 10^{-20} \ \mathrm{m}^{-1} \ \mathrm{s}^2 \ 10(\omega_L\tau_L^2)/(k_L''c) < 0.1$ holds.)

Moreover, introducing the retarded time $\eta = t - z/v$ with $v = \left(\dfrac{dk}{d\omega}\Big|_{\omega=\omega_L}\right)^{-1}$ being the group velocity at the carrier frequency $\omega_L$ and the space coordinate $\zeta = z$ in this moving frame, the relation (2.17) takes a form where first derivatives of the field strength only appear with respect to the spatial coordinate $\zeta$. Thus we arrive at

$$\frac{\partial}{\partial\zeta} \overline{E}(\eta,\zeta) - D_1 \overline{E}(\eta,\zeta) = 0 \qquad (2.20a)$$

where

$$D_1 = \frac{i}{2} k_L'' \frac{\partial^2}{\partial\eta^2} + c_L''' \frac{\partial^3}{\partial\eta^3} \qquad (2.20b)$$

is a differential operator describing pulse distortion by dispersion, and

$$c_L''' = \frac{1}{2k_Lc^2}\left(\varkappa_L' + \omega_L\varkappa_L'' + \frac{1}{6}\omega_L^2\varkappa_L'''\right) \qquad (2.20c)$$

In many cases one may, using the arguments given above, also neglect the second term in $D_1$. The remaining differential operator takes the form

$$D_L = \frac{i}{2} k_L'' \frac{\partial^2}{\partial\eta^2} \qquad (2.20d)$$

The general solution of this approximate equation for the pulse envelope with given entrance envelope $\overline{E}_0(\eta)$ is

$$\overline{E}(\eta,\zeta) = \frac{1}{2\sqrt{\frac{1}{2}i\pi k_L''\zeta}} \int_{-\infty}^{\infty} d\eta' \; \overline{E}_0(\eta') \exp\left\{i\,\frac{(\eta-\eta')^2}{2k_L''\zeta}\right\} \quad (2.20e)$$

## 2.5 Linear optical elements

Equation (2.15b) can be generalized for arbitrary linear elements, which are characterized by an optical transfer function $R(\omega)e^{-i\psi(\omega)}$, namely

$$\underset{\sim}{E}(\omega) = R(\omega)e^{-i\psi(\omega)} \underset{\sim}{E}_0(\omega) \quad (2.21a)$$

where in the case of a dispersive bulk medium of length z the amplitude response $R(\omega)$ and phase response $\psi(\omega)$ are given by

$$R(\omega) = e^{-k_i(\omega)z} \quad (2.21b)$$

$$\psi(\omega) = k_r(\omega)z = \frac{2\pi}{\lambda} n(\omega)z \quad (2.21c)$$

Obviously, $\psi(\omega)$ can also be interpreted as phase delay that an electromagnetic wave of frequency $\omega$ suffers while passing through a linear element having refractive index $n(\omega)$ and length z.

According to (2.20b), (2.21a), we can solve the pulse propagation problem in three straightforward steps:

(1) The incident pulse $E_0(t) = E(t,0)$ is Fourier-transformed with the result $\underset{\sim}{E}_0(\omega)$;
(2) the Fourier component $\underset{\sim}{E}_0(\omega)$ is multiplied by the propagator $\exp[-ikz]$ and $\underset{\sim}{E}(\omega,z)$ results;
(3) by Fourier back transformation of $\underset{\sim}{E}(\omega,z)$ we obtain the field strength of the transmitted light pulse $E(t,z)$.

If the preconditions for applying the carrier frequency and pulse envelope concept are satisfied, (2.21a) can be rewritten as

$$\overline{\underset{\sim}{E}}(\omega-\omega_L) = R(\omega)e^{-i\psi(\omega)} \overline{\underset{\sim}{E}}_0(\omega-\omega_L) \quad (2.22a)$$

Now we expand $\psi(\omega)$ around the carrier frequency $\omega_L$

$$\psi(\omega) = \sum_{n=0}^{\infty} b_n(\omega-\omega_L)^n \quad (2.22b)$$

where

$$b_n = \frac{1}{n!}\frac{d^n\psi}{d\omega^n}\bigg|_{\omega=\omega_L} = \frac{1}{n!}\psi_L^{(n)} \quad (2.22c)$$

Substituting (2.22b) in (2.22a) and back-transforming into the time domain give

$$\bar{E}(t) = e^{-ib_0} \int_{-\infty}^{\infty} \bar{E}_0(\omega - \omega_L) \, R(\omega) \, \exp[i(\omega - \omega_L)(t - b_1)]$$

$$\times \exp\left[ -i \sum_{n=2}^{\infty} b_n(\omega - \omega_L)^n \right] d\omega \qquad (2.22d)$$

It is obvious that $b_0 \neq 0$ causes a constant phase factor and $b_1 \neq 0$ results in a time delay, which is also called group delay. (In a dispersive bulk medium $b_1$ is related to the group velocity v through $b_1 = z \, k'(\omega) = z/v$). The frequency modulation of the pulses is affected only by terms with $n \geqslant 2$. As it is expected, (2.22d) represents the solution of equation (2.20a), if we assume loss-free propagation and expand $\psi$ up to $n = 3$.

It can be easily shown from (2.22d) for an unchirped pulse that terms with $b_2(\omega - \omega_L)^2 \neq 0$ (group velocity dispersion, GVD) affect modulus and phase whereas those with $b_3(\omega - \omega_L)^3 \neq 0$ change only the pulse envelope. Note that here and in what follows we will generally use the notation GVD for $b_2$ and $\psi_L^{(2)}$ keeping in mind that the GVD dv/dω of a dispersive sample of length l is related to $\psi$ by

$$\left. \frac{dv}{d\omega} \right|_{\omega = \omega_L} = -1 \frac{1}{(\psi_L^{(1)})^2} \psi_L^{(2)} \qquad (2.22e)$$

Obviously the sign of the GVD is determined exclusively by the sign of $\psi_L^{(2)}$ and $b_2$, respectively.

## 2.6  Basic principles of light pulse compression

As mentioned in the introduction, light pulse compression is achieved by two subsequent steps (Figure 2.2).

(1) The pulse is sent through a nonlinear sample, where a phase modulation is impressed on it which increases the pulse bandwidth. The features of this phase modulation depend on the actual physical mechanism leading to a nonlinear polarization. Various types of possible processes will be discussed in Section 4.

(2) The chirped pulse passes through a linear-optical sample that compensates for the phase modulation while the bandwidth remains unchanged. Consequently, the duration of the output pulse is determined by the width of the pulse spectrum behind the nonlinear sample

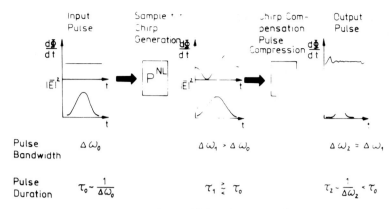

**Figure 2.2**  Scheme of light pulse compression.

and can therefore be significantly shorter than the duration of the input pulse.

To discuss the second step in more detail, we have written the chirped pulse in the frequency domain and multiplied it with the transfer function of the linear (lossless) element to get

$$a_2(\omega) = a_1(\omega)e^{i\phi(\omega)} e^{-i\psi(\omega)} \tag{2.23a}$$

In order to obtain an output pulse with maximum peak intensity (all spectral components in phase) $\psi(\omega)$ must be chosen by the requirement

$$\psi(\omega) = \phi(\omega) \tag{2.23b}$$

According to the discussion in Section 2.5, it is sufficient to require $\psi^{(n)}(\omega) = \phi^{(n)}(\omega)$ for $n \geqslant 2$. As in [2.5], [2.6] we will call an element of this kind an ideal filter or compressor. However, the ideal filter does not necessarily produce the best pulse compression because it can lead to undesired pulse shapes and satellites, although, it results automatically in the highest peak amplitude. The complete compensation of the spectral phase by means of the ideal filter reliably produce the shortest possible pulses with a clean profile only with linearly chirped Gaussian pulses.

In practice it is impossible to find the ideal filter for any arbitrarily given $\phi(\omega)$. Most of the linear elements used for pulse compression have a transfer function $\psi(\omega)$ where the second derivative $\psi_L^{(2)}$ dominates (quadratic compressor). The maximum pulse compression with such elements is approached if $\psi^{(2)}(\omega) \simeq \phi^{(2)}(\omega)$ and if $\phi(\omega)$ does not deviate from a parabolic profile too much [2.6], [2.5]. From the theoretical point of view an arbitrary phase function $\phi(\omega)$ can easily be fitted up to the n-th order (2.22b) if we use n linear elements $B_1, B_2, \ldots B_j, \ldots B_n$ in series exhibiting adjustable dispersion for which

$$\psi_L^{(m)} = \begin{cases} \dfrac{d^m\phi(\omega)}{d\omega^m} \bigg|_{\omega = \omega_L} & \text{for } m = j \\ 0 & \text{for } m \neq j \end{cases} \tag{2.24}$$

holds.

With respect to pulse compression we will deal with linear elements only that do not change the pulse energy (i.e. $R(\omega) = 1$). It should be noted that under certain conditions chirped pulses can also be compressed by simple spectral filtering which, however, is associated with energy loss. A similar effect can be observed, when only a certain part of the spectrum of a chirped pulse is amplified [2.7], [2.8]. Of course, for a shortening the spectrum of the chirped input pulse has to be broader and the inverse pulse duration has to be smaller than the gain profile. The gain profile provides in this manner an upper limit for the maximum achievable inverse pulse duration.

From a general point of view, pulse compression inside and outside a laser cavity is quite similar. However, the description of the intracavity process is more complicated in most cases because chirp generation and compensation can not be treated independently from the other pulse shaping mechanisms and certain steady state requirements must be fulfilled (Section 5).

## 3.  PRODUCTION AND COMPENSATION OF CHIRP IN LINEAR OPTICAL DEVICES

There are different types of linear elements which were used to produce, or more frequently, to compensate phase modulation for pulse compression. With this aim, one is generally interested to have elements which act only through their dispersive properties and leave the pulse energy unchanged, i.e. in terms of (2.21)

$$R(\omega) = 1 \tag{3.1a}$$

$$\psi(\omega) \neq \text{const.} \tag{3.1b}$$

Of course even in limited frequency ranges these requirements can only be approximately met due to the Kramers-Kronig relationship between $R(\omega)$ and $\psi(\omega)$ (see, e.g., [3.1a]). On the other hand, if a strong frequency dependence of $\psi(\omega)$ is desired, a near resonance effect can be advantageously used, but usually (3.1a) is violated. In the next three subsections we will describe various processes taking place in the most widely used linear optical components which have gained importance with respect to both the generation and to the extracavity compression of ultrashort light pulses.

## 3.1 Pulse shaping in traversing dispersive media far from resonances

This case is fully described by the linear wave equation (2.12) derived in Section 2.4 or by its representation in the frequency domain (2.15). To solve this problem it is sufficient to know $k(\omega)$. As an example, Figure 3.1 shows the absorption and dispersion as a function of the wavelength for fused silica which is one of the most commonly used bulk material for mirrors, prisms, fibers, etc., Obviously, the high transmittance and the dispersion profile justify to set $R(\omega) \simeq 1$ and to use only a few terms in the expansion of $\psi(\omega)$ (or $n(\omega)$) within a large frequency and wavelength range, respectively (0.4 $\mu$m–1.6 $\mu$m). To calculate the refractive index, a so-called Sellmeier equation for $n(\lambda)$ is often used whose coefficients are given e.g. in [3.1b] for different types of glass. For fused silica this equation reads, e.g.,

$$n^2(\lambda) = 1 + \frac{0.6961663\,\lambda^2}{\lambda^2 - 0.0684043^2} + \frac{-0.4079426\,\lambda^2}{\lambda^2 - 0.1162414^2}$$
$$+ \frac{0.89794\lambda^2}{\lambda^2 - 9.896161^2} \tag{3.2a}$$

According to (2.21c) the spectral phase response of a sample of length z is given by

$$\psi(\omega) = \frac{2\pi}{\lambda}\,n(\omega)z \tag{3.2b}$$

As an example we consider the propagation of a light pulse with carrier frequency $\omega_L$ and Gaussian-shaped pulse envelope at the entrance, the field strength and its Fourier transform are given by (2.8). Multiplication by the propagator and reverse Fourier transformation yields

$$E(t,z) = \frac{A_0}{4\sqrt{\pi}} \left\{ e^{i\Phi_0} \frac{1}{\sqrt{\gamma_0 - i\beta_0}} \int_{-\infty}^{\infty} d\omega \exp\left[ -\frac{(\omega - \omega_L)^2}{4(\gamma_0 - i\beta_0)} - ik(\omega)z + i\omega t \right] \right.$$

$$\left. + e^{-i\Phi_0} \frac{1}{\sqrt{\gamma_0 + i\beta_0}} \int_{-\infty}^{\infty} d\omega \exp\left[ -\frac{(\omega + \omega_L)^2}{4(\gamma_0 + i\beta_0)} + ik(\omega)z - i\omega t \right] \right\} \tag{3.2c}$$

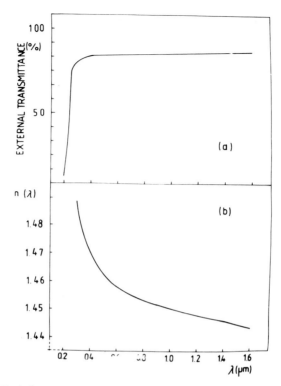

**Figure 3.1** Optical properties of fused silica as a function of the wavelength λ, a. external transmittance (note that about 8 percent losses from reflections at the input and excit plane have to be added to get the internal transmittance). b. refractive index.

For any given dispersion relation $k(\omega)$ of the sample, the field strength at the exit can be obtained by calculating these integrals. In approximation the integrals can be solved analytically by expanding $k(\omega)$ at $\omega = \omega_L$ and neglecting all terms higher than second order in $(\omega - \omega_L)$. The result is again a chirped Gaussian light pulse with envelope

$$\bar{E}(\eta,\zeta) = A_0 e^{i\Phi_0} \left[ 2k_L'' \, \xi_i \left( \frac{2\ln 2}{\tau_{L0}^2} - i\beta_0 \right) + 1 \right]^{-1/2} \exp\left[ -2\ln 2 \left( \frac{\eta}{\tau_L} \right)^2 \right]$$

$$\times \exp\left\{ -i\, 2\ln 2 \left( \frac{\eta}{\tau_L} \right)^2 \left[ \frac{4\ln 2 k_L'' \zeta}{\tau_{L0}^2} + \frac{\beta_0 \tau_{L0}^2}{2\ln 2} \, (1 + 2k_L'' \beta_0 \zeta) \right] \right\} \quad (3.3a)$$

where we have used the local coordinates $\eta$, $\zeta$. The pulse duration $\tau_L$ evolves as

$$\tau_L = \tau_L(\zeta) = \tau_{L0} \sqrt{1 + 2\beta_0 k_L'' L_\beta [1 - (1 - \zeta/L_\beta)^2]} \quad (3.3b)$$

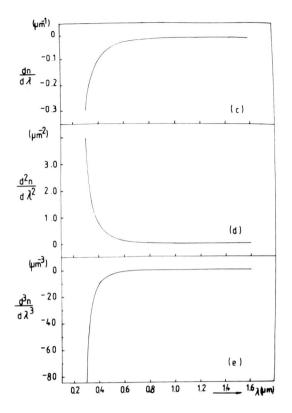

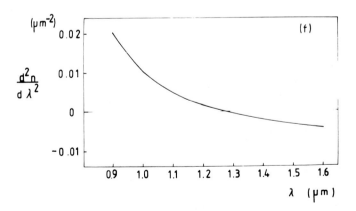

**Figure 3.1** c. first, d. second, e. third derivative of the refractive index with respect to λ. f. second derivative of the refractive index around the zero-dispersion wavelength. Note that $d^2\psi/d\omega^2 = L\lambda^3/2\pi c^2 \, d^2n/d\lambda^2$, where L is the length of the sample.

and the characteristic length $L_\beta$ is given by

$$L_\beta = - \frac{2\beta_0\, \tau_{L0}^4\, k_L''}{(4 \ln 2\; k_L'')^2 + (2\beta_0\, \tau_{L0}^2\, k_L'')^2} \qquad (3.3c)$$

For the chirp parameter $\beta$ we obtain

$$\beta(\zeta) = \frac{(2 \ln 2)^2\, k_L''(\zeta - L_\beta)}{\tau_L^2(\zeta)\cdot\tau_L^2(L_\beta)} \qquad (3.3d)$$

from which the instantaneous frequency

$$\omega(\eta,\zeta) = \omega_L + 2\beta(\zeta)\cdot\eta \qquad (3.3e)$$

is gained.

For $L_\beta > 0$, which requires $k_L''\beta_0 < 0$, the pulse is shortened on its path from $\zeta = 0$ to $\zeta = L_\beta$ (see Fig. 3.2a,b). At this point the pulse is bandwidth limited (chirp-free). For $\zeta > L_\beta$ the pulse duration increases monotonically and the pulse becomes phase modulated again. Bandwidth limited Gaussian input pulses double their duration at

$$\zeta = \frac{\sqrt{3}}{4 \ln 2}\, L_D \simeq 0.6\, L_D \qquad (3.4a)$$

where

$$L_D = \frac{\tau_{L0}^2}{k_L''} \qquad (3.4b)$$

denotes the so-called dispersion length of the material. Fig. 3.2a shows the dependence of the pulse duration and maximum intensity on the path length in the sample. Fig. 3.2b represents the pulse evolution in space and time and Fig. 3.2c demonstrate the generation of bandwidth limited pulses for down-chirped and up-chirped input pulses, which requires the appropriate sign of the GVD. Fig. 3.2d gives experimental results on the compression of down-chirped femtosecond pulses in SF5 glass [3.2] where $\tau_{L0} = 260\text{fs}$. The behavior depicted in Fig. 3.2a is also seen here.

We would also like to point out that these results can also be obtained with any other linear element for which it is justified to neglect terms with $n \geqslant 3$ in the expansion of the phase response (2.22b).

As a second example let us consider the propagation of an extremely short light pulse with Gaussian shaped field strength where we must leave the pulse envelope carrier frequency concept. The electric field of the input pulse is given by

$$E(t,0) = E_0(t) = \hat{E}_0\, e^{-2\ln2(t/\tau_{L0})^2} \qquad (3.5a)$$

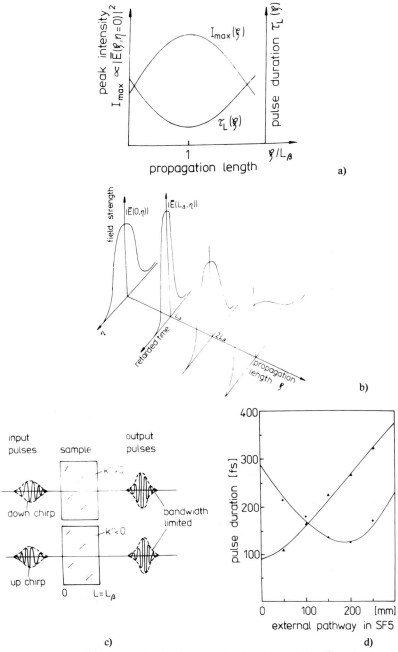

**Figure 3.2** Behavior of chirped pulses in traversing dispersive media ($\psi^{(2)}(\omega)$, $k''(\omega) \neq 0$), a. peak intensity and pulse duration versus the propagation length for $\beta_0 k'' < 0$, b. pulse envelope at certain positions in the dispersive sample, c. generation of bandwidth-limited pulses, d. experimental results on the propagation of down-chirped fs pulses ($\tau_{L0} = 260$ fs, $\cdot$) and bandwidth-limited pulses ($\tau_{L0} = 90$ fs, $\blacktriangle$) in SF5 glass (from [3.2]).

which reads in the frequency domain

$$\underline{E}_0(\omega) = \sqrt{\frac{\pi}{2 \ln 2}} \, \tau_{L0} \, \hat{E}_0 e^{-\frac{(\tau_{L0}\omega)^2}{8 \ln 2}} \tag{3.5b}$$

Note that $E_0(t)$ does not exhibit any periodical modulation due to a carrier frequency. Like before, we obtain at propagation length z

$$E(t,z) = \frac{1}{\sqrt{8\pi \ln 2}} \, \hat{E}_0 \int_{-\infty}^{\infty} d\omega \, \exp \left\{ -\frac{(\tau_{L0}\omega)^2}{8 \ln 2} + i\omega t - ik(\omega)z \right\} \tag{3.5c}$$

If we expand $k(\omega)$ again, now at $\omega = 0$, and neglect terms higher than the second order in $\omega$, a chirped Gaussian-shaped pulse is obtained. It again doubles its duration after having propagated in the sample the length.

$$z \simeq 0.6 \, L_D \tag{3.5d}$$

Let us now include some phase modulation

$$\Phi(t) = \alpha_0 t + \beta_0 t^2 \tag{3.6a}$$

For

$$\alpha_0 \tau_{L0} \ll 1 \text{ and } \beta_0 \tau_{L0}^2 \ll 1 \tag{3.6b}$$

the pulse also does not exhibit any oscillations of the field strength near its center. The Fourier amplitude of this input pulse is given by (2.8e), where $\omega_L$ has to be substituted by $\alpha_0$.

Such extremely short pulses have not been obtained yet for center frequencies $\omega_L$ ($\omega_L = \alpha_0$) in the visible region. In [3.3], however, the authors were successful in generating such light pulses in the far infrared ($\omega_0 \simeq 6 \cdot 10^{12} \text{ s}^{-1}$, $\lambda_0 \simeq 300 \text{ }\mu\text{m}$, $\tau_L \simeq 10^{-13} \text{ s}$) by optical rectification of visible light pulses, the FWHM of the envelope of which was on the order of 100fs (see Fig. 3.3). The propagation of such light is also given by (3.2) and (3.3).

## 3.2 Phase shaping through angular dispersion

### 3.2.1 General

Angular dispersion has been advantageously used for a long time to resolve spectra or for spectral filtering, utilizing the spatial distribution of the frequency components behind the dispersive element (e.g., prism, grating).

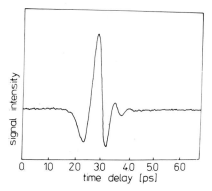

**Figure 3.3** Experimental observation of an electric-field waveform of approximately one cycle in duration in the Terahertz spectral range. The signal shown here corresponds to a convolution of the THz-signal with a 100 fs probe pulse (620 nm) from [3.3]).

There is a general relationship between angular dispersion and group velocity dispersion[1], which was first derived in [3.4] for the compression of chirped pulses with diffraction gratings. Later this concept was generalized and transferred to prisms and prism sequences [3.5].

From a general point of view, the diffraction can be treated by solving the Fresnel integrals which relate the angular dispersion and GVD in a rather complicated manner [3.4], [3.6], [3.7]. In order to discuss the main mechanisms we restrict ourselves to plane waves and use the advantages of a description by ray optics. First let us briefly introduce a simple relationship between angular dispersion and GVD in a more intuitive way.

We consider an input beam of mid frequency $\omega_L$ and spectral bandwidth $\Delta\omega$ which passes through a device D (Fig. 3.4) that causes angular dispersion (e.g. grating, prism). Now we calculate the optical path P from O to a plane $\overline{AB}$ which is perpendicular to the wave vector $\vec{k}_L$ and to the direction of the energy flow. This plane could be for instance the entrance surface of a light receiver at which the different spectral components of the wave superimposes. The phase which is essential for the superposition depends on the optical path from O to $\overline{AB}$. For light of frequency $\omega$ and wavevector k we find for the corresponding path

$$P(\omega) = \overline{OP} = \frac{1}{\cos \alpha} \qquad (3.7a)$$

---

[1] It should be noted that there are two effects leading to group velocity dispersion. On one hand, the angular dispersion alone caused GVD which will be considered in this section. On the other hand, angular dispersion is frequently accompanied by a spatial filtering and serves in this combination as a frequency filter. Such a filter is always a source of GVD if it is traversed off-reasonantly (see Section 3.3).

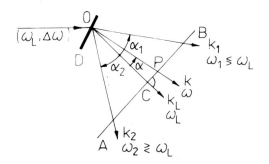

**Figure 3.4**   Scheme of angular dispersion.

where $l = P(\omega_L) = \overline{OC}$ and $\alpha = \alpha(\omega)$. The corresponding phase shift or phase response $\psi(\omega)$ is given by

$$\psi(\omega) = \frac{\omega}{c} P(\omega) = \frac{\omega l}{c \cdot \cos \alpha} \qquad (3.7b)$$

Thus, we have for the GVD at $\omega = \omega_L$ and at distance $l$ from the dispersive element

$$\psi_L^{(2)} = -\frac{\omega_L \, l}{c} \left( \frac{d\alpha}{d\omega} \bigg|_{\omega = \omega_L} \right)^2 \qquad (3.7c)$$

Obviously the sign of the GVD is always negative independent of the sign of the angular dispersion $(d\alpha/d\omega)$. If we consider a light pulse which is incident at 0 our detector at $\overline{AB}$ would measure a pulse changed in both modulus and phase due to the GVD. In the next two subsections we will describe grating and prism sequences which overcome the disadvantage of the divergence of the ray bundle and the distribution of the spectral components across a plane perpendicular to the energy flow. As we will see the basic mechanism leading to GVD is always angular dispersion where for a quantitative treatment only $d\alpha/d\omega$ has to be specified in (3.7c).

### 3.2.2   GVD of gratings

Similar to [3.4] we consider an input beam which is diffracted at the first grating $G_1$ at point A (Fig. 3.5). The grating $G_2$ which is parallel to $G_1$ redistributes the spectral components. Consequently the output beam is parallel with a frequency variation throughout the beam.

The optical path between A and an output wavefront $\overline{QQ}'$ is given by

$$P(\omega) = \overline{ACQ} = \frac{b}{\cos \alpha} (1 + \cos(\alpha + \beta)) \qquad (3.8a)$$

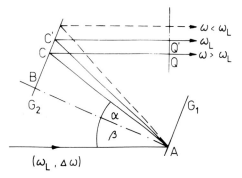

**Fig. 3.5** Scheme for calculating the GVD of a grating pair $G_1$, $G_2$. For convenience the reference wavefront is assumed to be in a position so that the extension of $\overline{QQ'}$ intersects $G_1$ at point A.

where $\beta$ is the angle of incidence, $\alpha$ is the diffraction angle and $b = \overline{AB}$ is the normal separation from $G_1$ to $G_2$. $\alpha$ and $\beta$ are related through the grating equation which reads for the first diffraction order

$$\sin \alpha + \sin \beta = \frac{2\pi c}{\omega d} \qquad (3.8b)$$

(d: grating constant).

In order to calculate the corresponding phase shift we have to account for a $2\pi$ phase jump at each ruling of $G_2$ [3.4] which would automatically follow from a wave optical treatment. Because only the relative number of phase jumps equaling $2\pi$ between C and C' is important, we count from B for simplification. Thus, we have

$$\psi(\omega) = \frac{\omega}{c} P(\omega) + \frac{b \tan \alpha}{d} \cdot 2\pi \qquad (3.8c)$$

Combining (3.8a–c) we find

$$\frac{d^2\psi}{d\omega^2} = - \frac{4\pi^2 cb}{\omega^3 d^2 r^{3/2}} \qquad (3.8d)$$

where $r = 1 - \left(\dfrac{2\pi c}{\omega d} - \sin \beta\right)^2$

and for the third derivative

$$\frac{d^3\psi}{d\omega^3} = - \frac{3}{\omega} \frac{d^2\psi}{d\omega^2}\left[1 + \frac{2\pi c\left(\dfrac{2\pi c}{\omega d} - \sin \beta\right)}{\omega d r}\right]. \qquad (3.8e)$$

Equation (3.8d) can obviously be rewritten as

$$\frac{d^2\psi}{d\omega^2} = - \frac{\omega b}{c \cdot \cos \alpha} \left(\frac{d\alpha}{d\omega}\right)^2 \tag{3.8f}$$

where $d\alpha/d\omega$ can be calculated from the grating equation (3.8b). If we remember that $b/\cos \alpha$ is the beam path between the two gratings, the analogy of (3.8d) and (3.7c) is obvious. This leads to some interesting conclusions.

The GVD is caused by the dispersion of the first grating; the second grating serves only as a recollimator which is in most cases necessary for further application, and to set a certain amount of GVD by adjusting the grating separation b. If the frequency variation over the beam cross section of the output beam is undesired, two grating pairs in series or a second passage through the same grating pair can be used, producing a beam collinear with the incident one and without transverse displacement of the spectral components.

To discuss the influence of higher order dispersion of gratings, we calculate the ratio of the second and third order term of the Taylor expansion for $\psi(\omega)$ (2.22b) using (3.8c), where $|\omega_L - \omega| \simeq \Delta\omega$ is assumed.

$$r_{32} = \left| \frac{b_3 \Delta\omega^3}{b_2 \Delta\omega^2} \right| \simeq \frac{\Delta\omega}{\omega_L} \left[ 1 + \frac{\lambda/d(\lambda/d - \sin \beta)}{1 - (\lambda/d - \sin \beta)^2} \right] \tag{3.8g}$$

Dealing with large bandwidths $\Delta\omega$, i.e., very short or extremely chirped pulses, one has to carefully prove whether $\psi^{(3)}(\omega)$ can be neglected or not. As an example we assume $\Delta\omega/\omega_L = 0.09$ (which corresponds approximately to 10fs Gaussian pulses at 600nm), $\lambda/d = 1.8$, $\beta = 30°$ and obtain $r_{32} = 0.2$. In most practiced cases grating pairs can be regarded to exhibit an ideal parabolic spectral phase response.

### 3.2.3   GVD of prisms

The same discussion as with GVD and angular dispersion of gratings holds for prisms and pairs of prisms (Fig. 3.6). However, besides angular dispersion, the material dispersion itself contributes to the GVD, which can approximately be taken into consideration by a mean (averaged over the beam cross section) glass path $l^2$. Whereas angular dispersion produces down chirp (3.7) the GVD of typical prism materials yields up-

---

[2] The averaging is only justified if the diameters of the input and dispersed beams are sufficiently small for the difference in the glass path over the beam cross section to be neglected. This effect alone was discussed in [3.8] and interpreted as the essential reason for GVD which is not valid. An exact calculation of both mechanisms shows, however, that the results derived in this section are applicable in most cases [3.7].

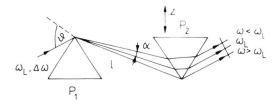

**Figure 3.6** Scheme for calculating the GVD of a pair of prisms $P_1$, $P_2$

chirp, which makes it possible to adjust the amount and the sign of the GVD simply by translating one or two prisms normally to their base while keeping their separation l constant [3.5]. The use of two pairs of prisms or a second passage can again be of advantage to avoid the lateral displacement of the spectral components in the output beam. To calculate the dispersion of the prism pair we need only to specify equation (3.7c) for the angular dispersion of prisms.

$$\left(\frac{d\alpha}{d\omega}\right)^2 = \left(\frac{d\alpha}{dn}\right)^2 \cdot \left(\frac{dn}{d\omega}\right)^2 \tag{3.9a}$$

where for Brewster prisms (i.e. $\vartheta$ corresponds to the Brewster angle for $\omega_L$) and minimum deviation $(d\alpha/dn)^2 \simeq 4$ holds [3.9]. Thus, we have

$$\frac{d^2\psi}{d\omega^2} = -\frac{\omega}{c} 1 \left(\frac{dn}{d\omega}\right)^2 \quad \times 4 \tag{3.9b}$$

where l is the distance between the two prisms. Note that the additional terms occurring in [3.5] vanish if the derivatives are calculated at the frequency for which $\beta$ (in our notation $\alpha$) approaches 0 (the minimum deviation is referred to the center frequency of the incident beam which has to serve as reference ray). In this sense the angular dispersion of prisms results also in the generation of down chirp. Of course, for the total GVD of prisms the contribution of a mean glass path has to be added in (3.9b) which is simply described by the second derivative of (2.21c).

As with gratings we want to estimate the effect of the next term in the Taylor expansion of $\psi(\omega)$. For this purpose we again calculate $r_{32}$ in the way described to find

$$r_{32} = \frac{1}{2} \frac{\Delta\omega}{\omega_L} \left[1 + 2\omega_L \left(\frac{dn}{d\omega}\bigg|_{\omega_L}\right)^{-1} \frac{d^2n}{d\omega^2}\bigg|_{\omega_L}\right] \tag{3.10}$$

which takes on a value of about 0.08 for the parameters of our previous example. This result supports the application of prism sequences as quadratic compressors i.e., with a dominant parabolic profile of the $\psi(\omega)$

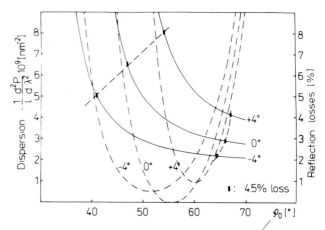

**Figure 3.7** GVD due to angular dispersion of a two-prism arrangement (SQ 1) at λ = 620 nm. The GVD (solid line) and the (reflection) losses (dash-dotted line) are calculated as a function of the angle of incidence for three values of the apex angle (0 corresponds to the Brewster angle of incidence for symmetrical beam path). $\psi_0$ corresponds to $\vartheta$ from Fig. 3.6, (from [3.10]).

curve. Moreover, prisms have the advantage of very small losses as opposed to the use of diffraction gratings. Note that $r_{32}$ can be significantly larger than estimated above if the prisms are not traversed under the minimum deviation angle.

In summarizing Fig. 3.7 shows the dispersion and the (reflection) losses of a two-prism arrangement as a function of the beam geometry and prism design (apex angle), where minimum deviation angle is provided.

## 3.3 Phase shaping by near resonant interaction

It is well-known from the fundamentals of optics that a near resonant interaction with a linear element not only distorts the amplitude of the electromagnetic field but also changes the phase in a characteristic manner. The amplitude and phase response are connected by the Kramers–Kronig relationship. Figure 3.8 shows the most simple case — the amplitude and phase response near a single transmission line as it occurs, for instance, with frequency filters.

Interferometric structures based on dielectric coatings have gained particular importance both in their use as reflecting mirrors [3.11-3.14] and in their use as specially designed interferometers [3.15], for adjusting a certain amount of dispersion [3.16-3.19].

In order to determine the response of a dielectric multilayer mirror, the waves reflected from each of the layer surfaces have to be superimposed

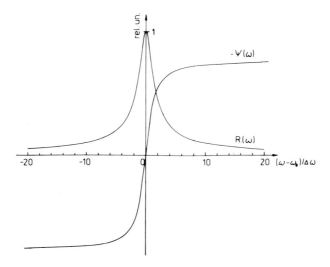

**Figure 3.8** Amplitude response $R(\omega)$ and phase response $\psi(\omega)$ for a complex Lorentzian shaped transmission line representing for example a frequency filter, where $R(\omega)e^{-i\psi(\omega)} = 1/(1 + i(\omega - \omega_0)/\Delta\omega)$.

where the layers differ in thickness and refractive index. This calculation can easily be performed by means of a matrix formalism [3.9] yielding $R(\omega)$ and $\psi(\omega)$. As an example, Fig. 3.9 shows the amplitude and phase response of a broadband high reflection mirror and a weak output coupler. Note, although they have similar reflection characteristics around the resonance ($\lambda = \lambda_0$), the dispersion differs greatly, which illustrates the significant influence of $R(\lambda)$ or $R(\omega)$ far from resonances for the dispersion curve near resonance.

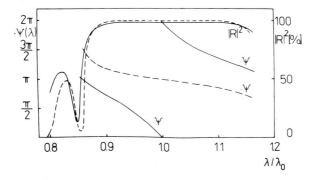

**Figure 3.9** Amplitude and phase response for a high reflection multilayer mirror (dashed line) and a weak output coupler (solid line), ($\lambda_0 = 2\pi c/\omega_0$: center wavelength) (from [3.11]).

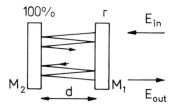

**Figure 3.10** Scheme of a Gires-Tournois interferometer .

From the various types of interferometers that, in principle, can be used to induce a phase modulation we want to briefly discuss the so-called Gires-Tournois interferometer [3.15]. This interferometer was first utilized to compress chirped pulses from a modelocked He-Ne laser [3.20] and it was later applied for intracavity femtosecond pulse compression by using dielectric multilayer structures [3.16–3.18].

The striking feature of this device is a very high and almost constant amplitude transmission while the GVD can be tuned from negative to positive values within several orders of magnitude. The device is essentially a Fabry-Perot interferometer where one mirror has an amplitude reflectivity r and the other is highly reflective (100%) (see Fig. 3.10).

The transfer function, according to (2.21a), can be obtained by adding the partial waves that leave the interferometer. In this manner one finds

$$\underset{\sim}{E}_{out}(\omega) = \left[ -r + \sum_{n=0}^{\infty} (1 - r^2) r^n e^{i(n+1)\beta} \right] \underset{\sim}{E}_{in}(\omega) \qquad (3.11a)$$

where $\beta = 2\omega d/c$ is the phase delay between two successive reflections at $M_1$. The geometric series in the brackets can easily be calculated and we find for the transfer function

$$R(\omega) e^{-i\psi(\omega)} = \frac{-r + e^{i\beta}}{1 - re^{i\beta}} \qquad (3.11b)$$

which finally yields [3.15]

$$R(\omega) = 1 \qquad (3.11c)$$

$$\tan \psi(\omega) = \frac{-(r^2 - 1) \sin \beta}{2r - (r^2 + 1) \cos \beta} \qquad (3.11d)$$

and

$$\frac{d^2\psi}{d\omega^2} = - \frac{2(1 - r^2) r \sin \beta}{(1 + r - 2r \cos \beta)^2} \left( \frac{d\beta}{d\omega} \right)^2 \qquad (3.11e)$$

$$\frac{d^3\psi}{d\omega^3} = -2(1 - r^2) r \frac{(1 + r^2) \cos \beta - 2r(1 + \sin^2\beta)}{(1 + r^2 - 2r \cos \beta)^3} \left( \frac{d\beta}{d\omega} \right)^3 \qquad (3.11f)$$

Obviously the dispersion can be controlled by changing the phase delay $\beta$.

To use these dispersive properties for femtosecond light pulses, the space between the two mirrors must be on the order of a few microns (corresponding to the geometrical length of the pulse). To overcome this obvious practical difficulty the use of dielectric coatings was proposed [3.17], [3.18] to construct Gires-Tournois interferometers. A corresponding device was built by placing a mirror $M_1$ (one $\lambda/4$ layer with a high refractive index) and a spacer d (k $\lambda/2$ layers) on a highly reflective dielectric multilayer mirror $M_2$ (a series of $\lambda/4$ layers), see Fig. 3.11. The dispersion can now be tuned by changing $\beta$ (here $\beta = \beta(d, \theta)$) through a variation of d or by changing the number of passes through the interferometers (i.e., the number of reflections at $M_{1,2}$ or similar arrangements).

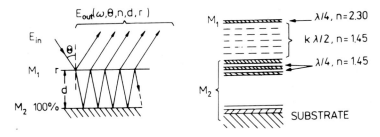

**Figure 3.11** Design of a Gires-Tournois interferometer using dielectric multilayers (from [3.17]).

It should be noted that (3.11) maintains its validity only if the dispersion of the multilayer system itself can be neglected. This, for example, can be assured for pulse carrier frequencies corresponding to the resonance frequency of the multilayer systems which requires a certain (constant)

**Table 3.1** GVD of various linear elements

| Linear element | Specification | | $\psi_L^{(2)}$ $(10^{-30}\ s^2)$ |
|---|---|---|---|
| Frequency Filter | $\omega_L = \omega_F - \frac{1}{8} \Delta\omega_F$ | $\Delta\omega_F = 52\,THz$ | 950 |
| | | $\Delta\omega_F = 520\,THz$ | 9.5 |
| | $\omega_L = \omega_F - \frac{1}{2} \Delta\omega_F$ | $\Delta\omega_F = 52\,THz$ | 740 |
| | | $\Delta\omega_F = 520\,THz$ | 7.4 |
| Glass | SQ1 | | $-52$ |
| (1 mm, 620 nm) | BK7 | | $-63$ |
| | SF5 | | $-180$ |
| Prism Pair | Material: SQ1 (620 nm) | | |
| | Prism Separation: 20 cm | | 175 |
| Grating Pair | 2000 mm$^{-1}$ | | |
| | Grating Separation: 20 cm | | 30 000 |
| Dielectric | $19 \times \lambda_0/4$ | $\omega_L/\omega_0 = 0.95$ | 26 |
| Multilayer | | $\omega_L/\omega_0 = 0.9$ | 210 |
| Mirror | | | |

value of $\theta$. If a change of $\theta$ is to control the dispersion as in [3.17], [3.18], the dispersion of the dielectric system contributes to the total dispersion of the device and modifies the transfer function.

To summarize the dispersive properties of the various linear elements discussed in this chapter, Table 3.1 presents the order of magnitude of $\psi^{(2)}(\omega)$ for some typical parameters.

## 4. GENERATION OF PHASE MODULATED PULSES BY NONLINEAR OPTICAL METHODS — PULSE COMPRESSION

So far we have discussed mechanisms for the production (or compensation) of phase modulation which do not change the spectral bandwidth if they are assumed to be loss-free. These processes are used in the second step of typical pulse compression experiments (cf. Fig. 2.2).

For the first step, as mentioned in Section 2, methods are neeeded by which the pulse spectrum becomes broadened with the instantaneous frequency exhibiting a simple dependence on time, where a linear increase or decrease of the frequency over the main part of the pulse is most desirable.

There is a variety of different nonlinear processes for impressing a phase modulation on light pulses. Refractive index changes in optical media which can be induced by active as well as passive methods have been advantageously used for this purpose. Active methods, e.g., based on the electro-optical effect, are limited by the rise time of the modulator and the synchronization which makes them applicable only for pulse durations above several 10 ps so far. Light-induced mechanisms, that modulate the phase and the frequency of the light pulse under consideration (signal pulse) with possible rise times less than $10^{-15}$ s overcome these restrictions. These mechanisms can be induced by the signal pulse itself. Then we speak of self-phase modulation (SPM), which, moreover, may offer the advantages of self-stabilization and self-controlling. On the other hand, the pump-induced phase modulation (PPM) can be used, where the nonlinear process leading to phase modulation of the signal pulse is induced by an additional pump pulse. Similarly to the active methods mentioned, PPM can advantageously be applied to control the parameters of the signal pulse.

The mechanisms behind light-induced refractive index changes can be divided into two main categories — nonresonant and near-resonant processes. Among the nonresonant mechanisms[3] are, e.g., the molecular

[3] Media are characterized as being nonresonant if their eigenfrequencies $\omega_{ij}$ are far from the center frequency $\omega_L$ of the light pulses, i.e., $|\omega_{ij} - \omega_L| \gg \Delta\omega_{ij}$, where $\Delta\omega_{ij}$ is to characterize the spectral width of the transition.

orientation Kerr effect in liquids, molecular electrostriction, thermal effects [4.1] and refractive index changes due to electronic nonlinearity. Only the latter distinguishes itself by a rise time $\leq 10^{-15}$ s which makes it applicable for pulse shaping on the femtosecond time scale. This electronic nonlinearity which causes an intensity dependent refractive index change has been investigated extensively in connection with self-focussing (see [4.2] and references therein), optical bistability [4.3] and more recently has found great importance in extracavity pulse compression [4.4–4.8], [4.28]. We will discuss its basic principles in the next subsection. Then we will discuss a nonlinear process — so-called parametric amplification — which enables to enhance the phase modulation of the incident pulse and to produce chirped pulses in other wavelength regions.

The near-resonance mechanisms, which·are mostly associated with the saturation of an amplifying or absorbing transition will then be dealt with in Section 4.2.

Finally, in section 4.3, the interplay between dispersion, nonresonant nonlinearities and resonant nonlinearities will be described with respect to pulse propagation in fibers under the presence of Raman processes.

For the description of the distortion of pulse modulus and phase under the conditions mentioned above we start from the wave equation (2.12), where now the nonlinear part of the polarization

$$P^{NL}(t,z) = \tfrac{1}{2} \overline{P}^{NL}(t,z)e^{i(\omega_L t - k_L z)} + c.c. \qquad (4.1)$$

has to be included. By using similar considerations as in section 2 we arrive at

$$\frac{\partial}{\partial \zeta} \overline{E}(\eta,\zeta) - D_L \overline{E}(\eta,\zeta) = \frac{i\mu_0}{2k_L} \left[ -\omega_L^2 \overline{P}^{NL}(\eta,\zeta) \right.$$
$$\left. + 2i\omega_L \frac{\partial}{\partial \eta} \overline{P}^{NL}(\eta,\zeta) + \frac{\partial^2}{\partial \eta^2} \overline{P}^{NL}(\eta,\zeta) \right] \qquad (4.2)$$

Within the frame of the SVEA the second and third term in the right hand side of (4.2) can be neglected. Depending on the processes to be considered the nonlinear polarization must be specified according to the nonlinear response of the medium.

## 4.1 Self-phase modulation in traversing dispersive optical samples with nonresonant nonlinearity for efficient pulse compression

### 4.1.1 *Derivation of the nonlinear Schroedinger equation (NLSE) and a generalized nonlinear Schroedinger equation (GNLSE)*

The chirp production in nonresonant optical media is described in most cases by treating the so-called nonlinear Schroedinger equation for the wave envelope $\overline{E}(t,z)$ of the light pulse (see, e.g., [4.9–4.11]). This equation will be derived and discussed in the following. Since we are interested in the limits of the method we also have to deal with more general relations and to study the effect of additional terms in a more general equation on the reshaping of light pulses.

The nonlinear polarization to be considered here can be written as

$$\overline{P}^{NL}(\eta,\zeta) = 2\epsilon_0 n[\bar{n}_2|\overline{E}(\eta,\zeta)|^2 + \tilde{n}_4|\overline{E}(\eta,\zeta)|^4 + \ldots]\overline{E}(\eta,\zeta) \quad (4.3a)$$

where

$$\bar{n}_2 \text{ and } \tilde{n}_4 = \bar{n}_4 + \frac{1}{2n}\bar{n}_2^2 \quad (4.3b)$$

are expansion coefficients of the nonlinear optical refractive index

$$n_{NL} = n + \bar{n}_2|\overline{E}|^2 + \bar{n}_4|\overline{E}|^4 + \ldots \quad (4.3c)$$

$\bar{n}_2 \neq 0$ results in the occurrence of the so-called Kerr-type nonlinearity.

It should be mentioned that in general $P^{NL}$ arises from a noninstantaneous response of the material. Here we need not take into account this memory effect since we restrict to samples which are almost lossfree and where, moreover, all frequencies are far from resonances. Modifications arising from the treatment of a more general equation (4.3a) will be discussed in connection with Raman processes in 4.3. The substitution of (4.3) into (4.2) yields

$$\frac{\partial}{\partial\zeta}\overline{E}(\eta,\zeta) - D_1\overline{E}(\eta,\zeta) =$$

$$-i[\tilde{\kappa}_2|\overline{E}(\eta,\zeta)|^2 + \tilde{\kappa}_4|\overline{E}(\eta,\zeta)|^4 + \ldots]\overline{E}(\eta,\zeta) \quad (4.4a)$$

where

$$\tilde{\kappa}_2 = \frac{\omega_L}{c}\bar{n}_2 \text{ and } \tilde{\kappa}_4 = \frac{\omega_L}{c}\tilde{n}_4 \quad (4.4b)$$

It should be mentioned here that the largest terms which have been neglected in deriving (4.4a) due to the SVEA and the assumption of instantaneous response are proportional to

$$\frac{2\tilde{\kappa}_2}{\omega_L}\frac{\partial}{\partial\eta}[|\overline{E}(\eta,\zeta)|^2\overline{E}(\eta,\zeta)] \quad (4.4c)$$

and

$$\omega_L\left[\frac{d}{d\omega}(\bar{n}_2 n)\right]_{\omega_L}\frac{\partial}{\partial\eta}\left\{\frac{1}{2\pi}\int_{-\infty}^{\infty}d\Omega|\underline{E}(\Omega)|^2\underline{E}(\Omega)e^{i\Omega\eta}\right\} \quad (4.4d)$$

The influence of such terms has been discussed, e.g., in [4.12–4.16].

Neglect of terms proportional to $c_L'''$ in the differential operator $D_1$ (2.20b) and $\tilde{\kappa}_4$ leads from (4.4a) to the NLSE

$$i \frac{\partial}{\partial \zeta} \bar{E}(\eta,\zeta) + \frac{1}{2} k_L'' \frac{\partial^2}{\partial \eta^2} \bar{E}(\eta,\zeta) - \tilde{\kappa}_2 |\bar{E}(\eta,\zeta)|^2 \bar{E}(\eta,\zeta) = 0 \quad (4.5a)$$

which describes pulse propagation under the simultaneous influence of group velocity dispersion and non-dispersive third order nonlinearity.

Substitution of $\epsilon = \sqrt{|\tilde{\kappa}_2/k_L''|} \ \bar{E}$ and $z' = \frac{\zeta}{2}|k_L''|$ in (4.5a) yields the NLSE in its standard form

$$i \frac{\partial \epsilon}{\partial z'} + (\text{sign } k_L'') \frac{\partial^2 \epsilon}{\partial \eta^2} - (\text{sign } \tilde{\kappa}_2) 2|\epsilon|^2 \epsilon = 0 \quad (4.5b)$$

which is often referred to in the literature.

A comprehensive tabulation of feasable bulk materials and their $\bar{n}_2$ which could be used for non-resonant phase modulation are found in [4.1]. In practice, however, glass fibers are mostly applied. Before solving (4.5) we therefore want to point out some aspects of the validity of the NLSE in fibers.

### 4.1.2 NLSE for single-mode fibers

Fibers have been proved to be well suited for the generation of phase modulation of optical pulses passing through them [4.17–4.19], [4.7],

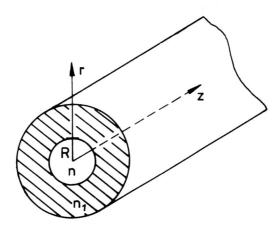

**Figure 4.1** Scheme of an optical fiber, where n ($n_1$) represents the refractive index of the core (cladding).

[4.10], [4.11]. The features which are advantageously used are the small losses (< 1 dB/km) and the relatively high field intensities over long propagation lengths due to the wave propagation as guided modes. The optical properties of the fiber core in which the pulses propagate are determined not only by the core material but also by that of the cladding (see Fig. 4.1) and the fiber geometry (see, e.g., [4.20]) resulting in an effective refractive index $n_{eff}$. As a consequence, the dispersion of a fiber differs from that of the corresponding bulk medium and can be controlled to some extent by the fiber geometry [4.19]. For instance it was possible to shift the zero-dispersion wavelength $\lambda_0$ (defined by $d^2 n_{eff}/d\lambda^2|_{\lambda_0} = 0$) from $\lambda_0 = 1.27$ $\mu$m to $\lambda_0 = 1.5$ $\mu$m [4.21] by changing the design of monomode fibers (fused silica as core material) or by proper doping materials [4.22].

The derivation of the NLSE concerning the guided modes is somewhat intricate for the general case. Here we only want to discuss briefly the modifications that are necessary for the equations from 4.1.1, to describe the pulse propagation in single mode fibers.

The nonlinearities are assumed to be small enough not to disturb the mode pattern determined by the linear wave equation and the boundary conditions. Furthermore, we neglect the longitudinal components of the electromagnetic waves. This approximation is justified for the fundamental mode in typical single mode fibers ($HE_{11}$-mode) [4.10], [4.11].

The starting point of the calculation is the wave equation (4.4a), where we now split the electric field into a radial (r) and a $\zeta$-dependent factor

$$\overline{E}(\eta,x,y,\zeta) = U(r)\,\overline{E}(\eta,\zeta) \qquad (4.6a)$$

Here U(r) is to satisfy the linear wave equation with the boundary conditions at $r = R$ which leads to the known mode profile. Dealing with single mode fibers we need to consider only the fundamental mode for which U is proportional to

$$U(r) \propto J_0(l_1 r/R) \qquad (4.6b)$$

where $J_0$ is the Bessel function of 0-th order and $l_1$ is its first zero.

Inserting (4.6a) in the wave equation, multiplying with rU(r) and integrating over the core cross section yields

$$i\frac{\partial}{\partial\zeta}\,\overline{E}(\eta,\zeta) - iD_l\overline{E}(\eta,\zeta) = \alpha_1\tilde{\kappa}_2\,|\overline{E}(\eta,\zeta)|^2\,\overline{E}(\eta,\zeta) +$$

$$+ \alpha_2\tilde{\kappa}_4\,|\overline{E}(\eta,\zeta)|^4\,\overline{E}(\eta,\zeta) \qquad (4.6c)$$

where the $\alpha_n = \left\{ 2\int_0^1 x dx (J_0(l_1 x))^{2n+1} \right\}/J_1^2(l_1)$ are constants on the order of 1 which remain to be calculated from the actual fiber geometry. For simplification we set $\alpha_n = 1$ and consequently have to refer $|\overline{E}|^2$, $|\overline{E}|^4$ to an effective core area and to replace $\tilde{\kappa}_4$ by $\tilde{\kappa}_4\,\dfrac{\alpha_2}{\alpha_1^2} = \tilde{\kappa}_4^{eff}$.

In this sense and by remembering the linear part n of $n_{NL}$ as an effective value $n_{eff}$ determined by the material and wave guide dispersion, the pulse propagation through the fiber is described by an equation analogous to (4.4a) and (4.5), respectively, which were derived for the bulk medium and to which we want to refer in what follows.

### 4.1.3  Solution method for the NLSE

The NLSE (4.5b) falls behind the class of nonlinear evolution equations for which the so-called inverse scattering method can be applied, see e.g. [4.23], [4.24]. Under the condition that the input pulse $\epsilon_0(\eta)$ vanishes sufficiently rapidly as $|\eta| \to \infty$, the initial value problem for the NLSE can be solved also by the use of this method (see, e.g., [4.24], [4.23]), which is to be sketched briefly following [4.25]. The input pulse $\epsilon_0(\eta)$ is considered as the potential in the equations of motion which read as follows for the NLSE

$$\frac{d}{d\eta} H_1 + \epsilon_0(\eta)H_2 = i\lambda H_1 \qquad (4.7a)$$

$$\frac{d}{d\eta} H_2 - \epsilon_0^*(\eta) H_1 = i\lambda H_2 \qquad (4.7b)$$

for the two-component function $\Phi = \begin{pmatrix} H_1(\eta,\lambda) \\ H_2(\eta,\lambda) \end{pmatrix}$, which depends on the eigenvalue $\lambda$. In the case to be discussed ($k_L'' \tilde{\kappa}_2 > 0$) the eigenvalue problem has no discrete solutions. Discrete eigenvalues appear with the opposite sign of $k_L'' \tilde{\kappa}_2$ and are characteristic for the occurrence of solitons (see 4.1.5.2). In view of the goal of the inverse scattering method we are only interested in a special solution of (4.7) which is fixed by its behavior for $\eta \to -\infty$, namely

$$\lim_{\eta \to -\infty} \Phi(\eta,\lambda) = \begin{pmatrix} 1 \\ 0 \end{pmatrix} e^{-i\lambda\eta} \qquad (4.8a)$$

Then for $\eta \to \infty$

$$\lim_{\eta \to \infty} \Phi(\eta,\lambda) = a(\lambda)\begin{pmatrix} 1 \\ 0 \end{pmatrix} e^{-i\lambda\eta} + b(\lambda) \begin{pmatrix} 0 \\ 1 \end{pmatrix} e^{i\lambda\eta} \qquad (4.8b)$$

holds, where

$$r(\lambda,0) = \frac{b(\lambda)}{a(\lambda)} \qquad (4.8c)$$

is called the reflection coefficient of the "input pulse potential". Following the general lines of the inverse scattering method (see, e.g., [4.26],

[4.27], [4.23]) the reflection coefficient of the (unknown!) "output pulse potential" $\epsilon(\eta, z')$ is easily obtained as

$$r(\lambda, z') = r(\lambda, 0)e^{4i\lambda^2 \zeta} \qquad (4.9)$$

From $r(\lambda, z')$ one can now gain the "potential" $\epsilon(\eta, z')$ itself by solving the corresponding linear Gelfand-Levitan-Marchenko integral equation, an approximate solution of which will be discussed in section 4.1.5.1.

### 4.1.4   Neglect of linear dispersion

Neglect of linear dispersion in (4.4a) ($D_1 = 0$) leads to a relation which has the following solution for a given input pulse $\overline{E}_0(\eta)$

$$\overline{E}(\eta, \zeta) = \overline{E}_0(\eta) \exp\{-i\zeta[\tilde{\kappa}_2 | \overline{E}_0(\eta)|^2 + \tilde{\kappa}_4 | \overline{E}_0(\eta)|^4 + \ldots]\} \quad (4.10a)$$

Obviously the nonresonant nonlinearity, for which the susceptibilities are real, changes only the phase of the light pulse, whereas the intensity profile remains unaffected. The time dependent phase and envelope (2.4b) at the output of a dispersionless nonlinear element are accordingly

$$A(\eta, \zeta) = A_0(\eta) \qquad (4.10b)$$

$$\Phi(\eta, \zeta) = \Phi_0(\eta) - [\tilde{\kappa}_2 | \overline{E}_0(\eta)|^2 + \tilde{\kappa}_4 | \overline{E}_0(\eta)|^4 + \ldots]\zeta \qquad (4.10c)$$

and the instantaneous frequency is given by

$$\omega(\eta) = \omega_L + \frac{d}{d\eta}\Phi_0 - [2\tilde{\kappa}_2 | \overline{E}_0(\eta)| + 4\tilde{\kappa}_4 | \overline{E}_0(\eta)|^3]\zeta \frac{d}{d\eta}A_0(\eta) \quad (4.10d)$$

If we have a chirpfree input pulse ($d^2\Phi_0/d\eta^2 = 0$) and $\tilde{\kappa}_4$ can be neglected the pulse is upchirped around its center while passing through the sample for $\tilde{\kappa}_2 > 0$, see Fig. 4.2a,b. Considering only the lowest order nonlinear optical effect, one may introduce the characteristic nonlinear optical interaction length

$$L_{NL} = \frac{1}{\tilde{\kappa}_2 A_{om}} \qquad (4.11)$$

where $A_{om}$ is the maximum of the envelope of the input pulse. $L_{NL}$ is a characteristic pathlength in a dispersionless nonlinear optical sample, after which the instantaneous phase is considerably changed (by 1 rad).

The calculation of nonlinear optical phase distortion with the neglect of group velocity dispersion is justified as long as the characteristic length $L_{NL}$ is small compared to the dispersion length $L_D$ given in (3.4b). For phase modulated input pulses of spectral with $\Delta\omega$ $L_D$ has to be substituted for by

$$L'_D = \frac{\pi^2}{(\Delta\omega)^2 \, k''_L} \tag{4.12}$$

By the use of high field strengths, large nonlinear optical phase shifts can indeed be produced with short sample lengths, for which group velocity dispersion can be neglected. At high field strength, however, the spectral phase shift is so complex, that it becomes impossible to find a loss-free linear optical filter for chirp compensation and pulse compression, which satisfies (2.23b).

To summarize the above discussion, Fig. 4.2 shows as an example the chirping of an unchirped sech input pulse in a nondispersive, nonlinear sample with $\tilde{\kappa}_2 \neq 0$ and $\tilde{\kappa}_4 = 0$. According to (4.10), the pulse envelope remains unaffected while the time-dependent phase varies as $\dfrac{d}{d\eta} A_0(\eta)$.

An ideal filter or compressor (Fig. 4.2c) compensates per definition (see discussion following (2.23b)) the spectral phase but does not lead necessarily to a perfect pulse compression. The main pulse is shorter than the input pulse, but is accompanied by undesirable satellites. The nonlinear behavior of $\Phi(\eta)$ is particularly unfavorable for compression with a quadratic compressor as can be seen from Figs. 4.3d, e. Fig. 4.3d shows the result after passage through a quadratic compressor that compensates for $\phi(\omega)$ exactly at the center frequency of the spectrum leading to an extreme pulse fragmentation and temporal spreading. In Fig. 4.3e, the pulse was compressed in an element the transfer function of which fits $\phi(\omega)$ approximately over a broad frequency range where most of two spectral intensity is located.

### 4.1.5  Simultaneous action of nonlinearity and dispersion — the key for efficient pulse compression

Obviously it is very difficult to compensate for the phase modulation occurring in dispersionless samples for pulse compression. However, it was found that the interplay between nonlinearity $\tilde{\kappa}_2$ and dispersion $k''_L$ as described by the NLSE for the simplest case may lead either to simple dependencies of the instantaneous frequency on time for $k''_L \tilde{\kappa}_2 > 0$ [4.29] or to soliton propagation phenomena for $k''_L \tilde{\kappa}_2 < 0$ [4.30], [4.9], [4.24]. Both mechanisms have been advantageously used for compression and will be discussed in detail.

### 4.1.5.1  Pulse shaping for GVD and nonlinearity having equal signs — Extracavity pulse compression. In third-order nonlinear optical media

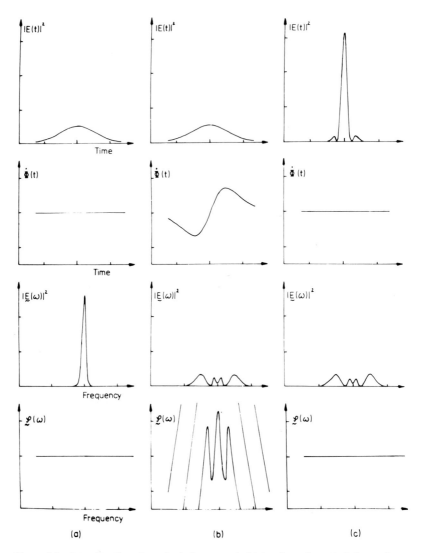

**Figure 4.2**  Intensity, time-dependent phase, spectral intensity and spectral phase of, a. a sech² chirpfree input pulse, b. the pulse from a) at the output of a nonlinear, nondispersive sample, c. the pulse from b) behind an ideal compressor, d. the pulse from b) behind a quadratic compressor that compensates $\phi(\omega)$ exactly only at the center frequency of the spectrum, e. the pulse from b) behind a quadratic compressor having a transfer function that approximately fits $\phi(\omega)$ in a frequency range where most of the spectral intensity occurs.

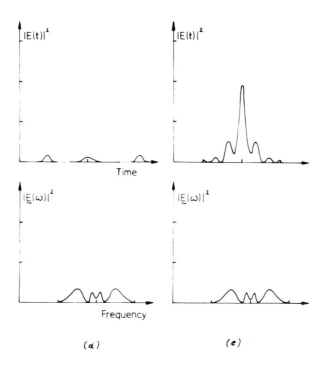

**Figure 4.2** *contd*

with $\tilde{\kappa}_2 > 0$, which is the usual sign for electronic contributions to non-linearity, light pulses become upchirped. For positive GVD ($k_L'' > 0$) the low-frequency leading edge of the pulse travels faster than its high-frequency trailing edge. ($k_L'' > 0$ and $\tilde{\kappa}_2 > 0$ is met for example in fused silica single mode fibers for wavelengths $\lambda_L$ below the zero-dispersion wavelength $\lambda_0 \simeq 1.3 \mu m$). Thus, in passing through the medium, the frequency sweep increases, but so does the pulse duration. At first glance this might seem to be a disadvantage. Since, however, the instantaneous frequency becomes an almost linear function of time, these spectrally broadened pulses can easily be compressed nearly until the bandwidth limit.

Chirp production under such conditions was first calculated in [4.29] by numerical integration of the nonlinear Schroedinger equation. It was found that an unchirped sech²-input pulse evolves into an almost rectangular output pulse, the instantaneous frequency of which increases linearly with time.

More extensive numerical studies of such pulse propagation problems were performed in [4.31]. Figure 4.3 gives typical results of the evolution

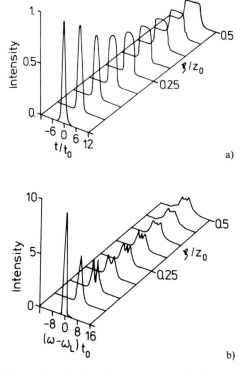

**Figure 4.3** Evolution of intensity a) and power spectrum b) of the pulse against the path length in the dispersive sample (from [4.31]). Note that $z_0 \simeq 0.5 L_D$ and $t_0 = \tau_L/1.76$.

of the instantaneous intensity and of the power spectrum of the pulse against the path length in the nonlinear optical medium. The pulses are seen to spread and develop a profile with steep leading and trailing edges. In these regions the rapidly varying intensity causes rapid variation of the instantaneous phase, and thus the generation of new frequencies appears primarily there.

Figure 4.4 shows the instantaneous frequency and the spectral phase of the light pulse at $\zeta = 0.25\ L_D$. Obviously the output pulse exhibits an almost linear increase of the instantaneous frequency and nearly rectangular intensity profile (cf. Fig. 4.3). The spectral phase varies in good approximation quadratically with frequency.

*Analytical approach.* To discuss the functional dependence of the parameters of such chirped output pulses versus parameters characterizing the

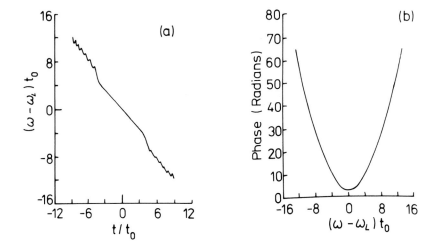

**Figure 4.4** Instantaneous frequency (a) and spectral phase (b) at $\zeta = 0.25L_D$ (from [4.31]). Note $t_0 = \tau_L/1.76$.

experimental situation, the nonlinear Schroedinger equation was treated by approximative analytical means in [4.25] following the main lines of the inverse scattering method mentioned in Section 4.1.3.

According to the numerical solution [4.29], [4.31] discussed above, the pulse exhibits a rectangular shape and the frequency increases almost linearly with time after the passage through a certain length of the fiber, provided that the pulse and fiber parameters satisfy the relation

$$L_D \gg L_{NL} \tag{4.13}$$

Assuming modulus and phase to vary as

$$\overline{E}_r(\eta) = \begin{cases} A_{rm}\, e^{i\beta\eta^2} & \text{for } |\eta| \leq \dfrac{\tau_{Lr}}{2} \\[2ex] 0 & \text{for } |\eta| > \dfrac{\tau_{Lr}}{2} \end{cases} \tag{4.14}$$

the reflection coefficient $r(\lambda,z')$ see (4.9) can be calculated exactly. The solution contains the unknown pulse parameters, however, which are related to each other in a rather complicated manner. If we further assume that the fiber is sufficiently long

$$\zeta_F \gg 0.5\sqrt{L_D\, L_{NL}} \tag{4.15}$$

which implies that due to the broadening the pulse would be affected mainly by the dispersion for $\zeta > \zeta_F$, r can be evaluated in approximation. Utilizing the following conservation laws of the NLSE

$$\frac{\partial}{\partial \zeta} \int_{-\infty}^{\infty} |\bar{E}|^2 \, d\eta = 0 \tag{4.16a}$$

$$\frac{\partial}{\partial \zeta} \int_{-\infty}^{\infty} \left( \left| \frac{\partial^2 \bar{E}}{\partial \eta^2} \right|^2 + |\bar{E}|^4 \right) d\eta = 0 \tag{4.16b}$$

where (4.16a) represents the conservation of energy, the parameters of the pulse behind the fiber of length $\zeta_F$ were found to be approximately [4.25]

$$\tau_{Lr} \simeq 2.9 \frac{\zeta_F}{\sqrt{L_D L_{NL}}} \tau_{L0} \tag{4.17a}$$

$$A_{rm} \simeq 0.6 \sqrt[4]{\frac{L_D L_{NL}}{\zeta_F^2}} A_{om} \tag{4.17b}$$

In agreement with the discussion above the spectral width of the pulse

$$\Delta \omega_r \simeq 1.4 \frac{1}{\tau_{L0}} \sqrt{\frac{L_D}{L_{NL}}} \tag{4.18a}$$

does not change any more if $\zeta_F$ is large enough. (The intensity is too small to generate noticeable nonlinear phase modulation.)

The chirp parameter

$$\beta = \frac{\Delta \omega_r}{2} \tau_{Lr} \tag{4.18b}$$

varies with $\zeta_F$ through $\tau_{Lr}$.

The fiber length at which 95% of the maximum achievable spectral width $\Delta \omega_r$ is obtained is given by

$$\zeta_F = 5.6 \sqrt{L_D L_{NL}} \tag{4.18c}$$

Now we calculate the compression of the pulse (4.14) in a linear element with GVD. For simplicity we rewrite the right side of the equation (2.22d) in the time domain, i.e., as the convolution of the pulse amplitude $\bar{E}_r(\eta)$ and the response function of the GVD element (for example, $b_n = 0$ in (2.22d) for $n > 2$) to get

$$\bar{E}(\eta) = A(\eta) e^{i\Phi(\eta)} = \frac{1}{\sqrt{-8\pi i b_2}} A_{rm} \int_{-\frac{\tau_{Lr}}{2}}^{\frac{\tau_{Lr}}{2}} d\eta' e^{i\beta\eta'^2} \exp\left\{ -i \frac{(\eta - \eta')^2}{4b_2} \right\}$$

$$\tag{4.19}$$

(Note that for a dispersive bulk medium of length $\zeta$ $b_2 = -0.5\,\zeta\,k_L''$ holds.)

Generally one is interested to have chirp-free output pulses of shortest possible duration where possible satellites are negligibly weak. The evaluation of equation (4.19) with the requirement of $d\Phi/d\eta = 0$ in the center of the output pulse yields

$$b_2 = \left[ 4\beta \left( 1 + \frac{22.5}{\beta^2 \tau_{Lr}^4} \right) \right]^{-1} \tag{4.20a}$$

for the GVD parameter $b_2$ [4.32].

For large fiber length $\beta\tau_{Lr}^2 \gg 10$ holds and $b_2$ simplifies to

$$b_2 = \frac{1}{4\beta} \tag{4.20b}$$

For this limiting case the integral (4.19) can be calculated analytically with the result

$$A(\eta)e^{i\Phi(\eta)} = \frac{A_{rm}}{\sqrt{8\,\pi\,b_2}}\;\frac{1}{b_2\eta}\,\sin\left(\frac{\tau_{Lr}}{4b_2}\,\eta\right)\exp\left\{-i\left(\frac{\eta^2}{4b_2} - \frac{\pi}{4}\right)\right\} \tag{4.21}$$

Equation (4.21) is now used to estimate the resulting pulse duration if a fiber of finite length $\zeta_F$ (4.18c) is used. ($\zeta_F \to \infty$, which was the exact precondition for deriving (4.21) would automatically lead to the impracticable requirement $b_2 \to \infty$). Combining (4.17), (4.18b,c) and (4.20b) we can calculate in approximation the duration of the compressed pulse from (4.21) and find

$$S = \frac{\tau_{L0}}{\tau_L} \simeq 0.5\,\sqrt{\frac{L_D}{L_{NL}}} \tag{4.22}$$

for the compression factor [4.25].

The structure of (4.21) already indicates some problems associated with the pulse compression in this way. On one hand, the output pulse is chirped with the chirp parameter $\tilde{\beta} = -\beta\;(= -1/4b_2)$ which is undesireable in some cases. On the other hand, the pulse exhibits satellites, though the ratio of the peak intensity to the first satellite is about 20. According to (4.20a), the output pulse is expected to be unchirped at $\eta = 0$ at values of the GVD parameter $b_2$ somewhat smaller than $1/4\beta$. This should also lead to a somewhat smaller pulse duration [4.32] but it does not necessarily mean an even better pulse quality.

*Numerical evaluation.* Numerical calculations were performed [4.31] which take into account the actual shape of pulse modulus and phase at the output of a fiber of finite length. Moreover, different values of $b_2$ were considered to find an optimum compression. It could be shown that the requirements for producing chirp-free output pulses of the shortest achievable duration and best quality can not be satisfied simultaneously,

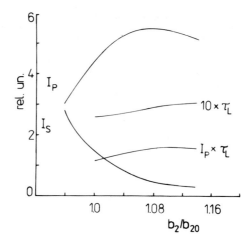

**Figure 4.5** Various parameters of the compressed pulse as a function of the GVD parameter $b_2$ normalized to $b_{20}$, where $b_{20}$ corresponds to the best fit of the spectral phase of the pulse at the output of the fiber. ($I_p$: peak intensity of the main pulse, $I_s$: ratio of the peak intensity of the first sidelobe and the input pulse) (from [4.31]).

see Fig. 4.5. Nevertheless, an optimum compression can be defined at values of $b_2$ that lead to the pulse with maximum peak intensity. Figure 4.6 shows the compression factors as a function of the fiber length under optimum compression.

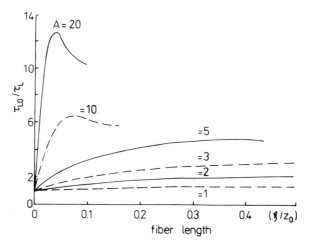

**Figure 4.6** Pulse compression with an optimum quadratic compressor as a function of the fiber length for different intensities of the input pulse, (from [4.31]). Note, $z_0 \simeq 0.5 L_D$ and $A \simeq 0.6\sqrt{L_D/L_{NL}}$.

**Table 4.1**  Results of typical pulse compression experiments

| Ref. | Input Pulse | | Fiber | | Output Pulse | | Compression Factor | |
|---|---|---|---|---|---|---|---|---|
|  | $\tau_{L0}$ ps | Peak Power kW | Core μm | Length m | $\tau_L$ ps | Energy nJ | Exp. | Theor. (max.) |
| [1.19] | 0.09 | 7 | 3.3 | 0.15 | 0.03 | 0.3 | 3 | 3 |
| [4.46] | 0.065 | 100 |  | 0.008 | 0.016 | 2 | 4 | 7 |
| [4.29] | 6 | 0.1 | 4 | 30 | 0.6 | 0.3 | 10 | 11 |
| [4.37] | 5 | 0.6 | 4 | 30 | 0.45 | 1 | 12 | 28 |
| [4.7] | 5.5 | 0.01 | 4 | 70 | 1.5 | 0.02 | 3.7 | 3 |
| [4.34] | 5 | 40 | 50* | 0.4 | 0.7 | $10^5$ | 7 | 60 |
| [4.65] | 100 | 0.1 | 8 | 2000 | 2 |  | 50 | 83 |
| [4.48a] | 0.04 | 120 | 4 | 0.007 | 0.008 |  | 5 | 4 |
| [4.48b] | 0.05 | 320 | 4 | 0.009 | 0.006 | 12 | 8 | 8 |

\* graded core index fiber

Obviously, depending on the duration and energy of the input pulse, there is an optimum fiber length at which the pulse compression is most effective and which from the numerical results could be estimated as [4.31]

$$\zeta_{opt} \simeq 1.4 \sqrt{L_D L_{NL}} \qquad (4.23a)$$

The corresponding compression factor can be written as

$$S_{opt} \simeq 0.37 \sqrt{\frac{L_D}{L_{NL}}} \simeq 0.37 \, \tau_{L0} \, A_{om} \sqrt{\frac{\tilde{\kappa}_2}{k_L''}} \qquad (4.23b)$$

and the pulse duration achieved finally is

$$\tau_L \simeq 2.7 \sqrt{\frac{k_L''}{\tilde{\kappa}_2 \, A_{om}^2}} \qquad (4.23c)$$

(The overestimation of S in the analytical approach (4.22) results from the assumption of an ideal rectangular pulse with linear chirp for $\zeta \to \infty$.)

Table 4.1 shows results of typical compression experiments and compares the compression factors with the values calculated from (4.23b). The corresponding experimental arrangements are similar to the one sketched in Fig. 4.7. The optical pulse is focussed through a microscope

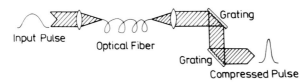

Input Pulse  Optical Fiber  Grating  Grating  Compressed Pulse

**Figure 4.7**  Basic experimental setup for light pulse compression using an optical fiber and grating pair.

objective into a single mode fiber. After recollimation the chirped and
temporarily broadened pulse passes through a grating pair for chirp compensation and pulse compression. It should be noted that instead of the
grating pair, prism sequences (Sect. 3) were successfully applied [4.33] for
extracavity pulse compression. In [4.34] a graded-core index fiber was
used, which, due to a relatively large core diameter, allowed higher pulse
energies. In addition, an amplifier stage was placed between the fiber and
the grating to reach the high output pulse energy. The pulses with a
duration of 6fs [4.48], which are the shortest light pulses achieved so far,
were obtained after suitable adjustment of two dispersion orders of the
linear element independently (see also next subsection).

From equation (4.23b) it is evident that samples for which $\tilde{\kappa}_2/k''$ is very
high are advantageous. Note that the optimum length of the nonlinear
sample, which yields almost maximum compression is proportional to

$$\sqrt{L_D L_{NL}} = \frac{\tau_{L_0}/A_{om}}{(\tilde{\kappa}_2 k''_L)^{1/2}} \qquad (4.24)$$

so the sample must be long if $k''_L$ is small. Furthermore, when $k''_L$ is very
small and the sample very long, we must take into account further
dispersion terms in equation (4.4). Moreover, the contribution of the
stimulated Raman scattering mostly undesired is proportional to the
product of fiber length and pulse intensity (for a discussion, see 4.3).
Therefore, any practical device should be fabricated from a material
having a large value of the nonlinear coupling parameter; semiconductors
and certain polymers having conjugated chains should be suitable [4.1].
However, so far it has proved impossible to confine optical beams of small
diameter over large path lengths in such materials because single-mode
optical fibers of such materials have yet to be produced.

*Limits of pulse compression.* There are several reasons why pulse
compression is limited in a nonlinear material, and some are now
discussed in turn. First, we consider the finite range of optical transmission of the sample, the spectral range of the phase-modulated pulse,
and thus the minimum achievable pulse duration. With typical single-
mode fibers used in pulse compression experiments, the transmission is
high over the frequency range from the near ultraviolet ($\nu_{UV}$) through the
visible, to the near-infrared ($\nu_{IR}$). Indeed, it is possible to generate spectral
continua by self-phase modulation with a spectral range of more than two
octaves from $\nu_{IR}$ to $\nu_{UV}$. Hence, there is a lower limit of the minimum pulse
duration, given by

$$(\tau_L)_{min} \simeq \frac{1}{\nu_{UV} - \nu_{IR}} \qquad (4.25)$$

This limit is considerably less than the wave period T of the wave having carrier frequency $\nu_L = \frac{1}{2}(\nu_{UV} + \nu_{IR})$. At present, however, there is no suitable method of compensating the complicated phase modulation of such extremely frequency-broadened pulses in order to compress them. Note that the self-phase modulation and compression cannot be described by the relationships derived here since the various assumptions used in deriving them are no longer satisfied. This means that the full wave equation has to be solved.

Secondly, we consider the limit of pulse duration achievable by using the experimental and theoretical methods described which is closely related to the range of validity of the NLSE (4.5). For a nonlinear sample the minimum pulse duration is determined by the maximum field strength $A_{om}$ of the input pulse (4.23c), which in turn is determined by the damage threshold of the material. In most cases, however, the strong convergence of the equation of nonlinear polarization in powers of field strength (4.3a) is already violated far below this threshold, and hence the simultaneous effect of many orders has to be taken into account. For very short samples the evolution of the temporal phase is given by equation (4.10c), to which in the simplest case further nonlinear terms of higher order in the expansion of the refractive index with respect to the intensity have to be added. (Terms describing the Raman effect which represent nonlinearities on the same order as $\tilde{\kappa}_2$ and which become important after longer propagation lengths will be considered in section 4.3.)

We now use equation (4.10c) to estimate the range of applicability of the nonlinear Schroedinger equation. Higher-order nonlinearities in equation (4.10c) can be neglected if

$$\zeta \tilde{\kappa}_4 |A_{om}|^4 \leqslant 1 \qquad (4.26)$$

The use of equation (4.10c) would be justified in this case because the higher-order nonlinearities affect the light pulse mostly in the initial part of the pathlength as long as the pulse is not spread by dispersion and its maximum field strength is approximately equal to the initial value $A_{om}$. Therefore, we consider pulses which are of constant maximum field strength between $\zeta = 0$ and $\zeta = \frac{1}{2}\zeta_{opt}$, where the optimum pathlength $\zeta_{opt}$ in a third order nonlinear sample is given by (4.23a). Since at $\zeta = \zeta_{opt}$ the maximum field strength already deviates by 20 percent from its value at $\zeta = 0$, we overestimate the effect of higher-order nonlinearities in this range. For $\zeta$ above $\frac{1}{2}\zeta_{opt}$ we neglect the higher-order nonlinear terms; this second inaccuracy partially compensates for the first one. Thus, we substitute $\zeta$ in (4.26) by $0.7(L_D L_{NL})^{1/2}$ and calculate the highest useable field strength, for (4.26) to remain valid. Insertion of this pulse amplitude into equation (4.23c) yields for the duration of the compressed pulse

$$(\tau_L)_{min} \simeq 8 \sqrt[3]{k_L'' \, \tau_{L0} \, \tilde{\kappa}_4/\tilde{\kappa}_2^2} \qquad (4.27a)$$

For $\lambda_L \simeq 0.6\mu m$ and typical glass single-mode fibers ($k_L'' \simeq 6.5 \times 10^{-26}$ s$^2$ m$^{-1}$, $\tilde{\kappa}_2 \simeq 6.5 \times 10^{-16}$ mV$^{-2}$, $\tilde{\kappa}_4 \simeq 10^{-36}$ m$^3$V$^{-4}$) one obtains from equation (4.27a)

$$\left(\frac{\tau_L}{sec}\right)_{min} \simeq 10^{-10} \left(\frac{\tau_{L0}}{sec}\right)^{1/3} \qquad (4.27b)$$

The structure of equations (4.27a) and (4.27b) suggests that several compression stages should be used to obtain successively shorter input pulses for each stage. Two-stage compression devices have already been used for long input pulses [4.37], see Fig. 4.9, but the fact that the minimum achievable pulse duration depends only on the third root of the input pulse duration limits the efficiency of such devices.

With input pulse durations of about 50 fs, equation (4.27b) gives a value of about 5 fs for the minimum achievable duration of the output pulses. With Fourier-limited output pulses at $\lambda_L = 0.6$ $\mu$m this duration corresponds to a bandwidth of about $6 \times 10^{13}$ Hz or 90 nm.

In extreme cases the time derivative of field strength and polarization $|\partial \bar{x}|\partial \eta|$ can attain values of about $\frac{1}{2}\omega_L|\bar{x}|$. Dealing with such short input pulses the term (4.4c) has to be considered independent of how large the input intensity is. At moderate intensities ($\tilde{\kappa}_4 \, \zeta \, |A_{om}|^4 \ll 1$) the starting equation reads

$$\frac{\partial}{\partial \zeta} \bar{E}(\eta,\zeta) - \frac{1}{2} ik_L'' \frac{\partial^2}{\partial \eta^2} \bar{E}(\eta,\zeta) = -i\tilde{\kappa}_2 |\bar{E}(\eta,\zeta)|^2 \bar{E}(\eta,\zeta)$$

$$- 2 \frac{\tilde{\kappa}_2}{\omega_L} \frac{\partial}{\partial \eta} |\bar{E}(\eta,\zeta)|^2 \bar{E}(\eta,\zeta) \qquad (4.28)$$

where the ratio of the nonlinear terms at the right side is on the order of $\omega_L \tau_{L0}$. In [4.35] the chirping described by (4.28) and the subsequent compression with a quadratic compressor was calculated. It could be shown that the spectral intensity as well as the instantaneous frequency develop an asymmetrical behavior with respect to the center frequency and pulse center, respectively, as the pulse propagates through the sample. The result is a smaller value of the maximum achievable compression (see Fig. 4.8) as compared with the case where the consideration of the last term in (4.28) is not necessary. If the term, representing the third order dispersion $c_L''' \frac{\partial^3}{\partial \eta^3} \bar{E}(\eta,\zeta)$ is additionally introduced in (4.28), a strong asymmetrical intensity profile at the output of the fiber also results [4.36].

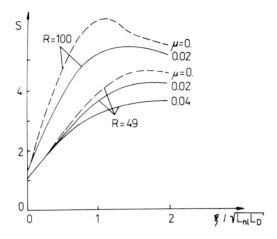

**Figure 4.8** Compression factors versus the normalized length $\zeta/\sqrt{L_D L_{NL}}$ for two values of R and different $\mu$, where $R \propto L_D/L_{NL}$ and $\mu = 2/(\omega_L \tau_{L0})$. (Note that $\mu$ characterizes the influence of the last term in (4.28) which is neglected when considering the NLSE) (from [4.35]).

*Compression experiments.* Presently the ultimate temporal resolution of ultrafast processes is determined by the laser pulse widths. The development of optical pulse compressors was stimulated by the search for even shorter pulses than those available directly from (passively mode-locked dye) lasers ($\simeq$ 20 fs-100 fs). Pulse compression techniques also provided the possibility to produce ps and sub ps pulses starting from other types of lasers and hence in various wavelength ranges, e.g., [4.38-4.44]. Dealing with extremely short pulses, i.e., starting with the fs light pulses from dye lasers extracavity pulse compression is often accompanied by the need for amplified pulses which is also frequently a precondition for the applications.

One of the simplest techniques to increase the pulse intensity from dye lasers is cavity dumping [4.45], which, combined with synchronous pumping, yields pulse intensities that are high enough for extracavity compression. One of the earliest and most successfully applied compressors is shown in Fig. 4.9. The 5.9 ps pulses from a cavity dumped, synchronously pumped dye laser were compressed in a two-stage arrangement by a factor of 65. In addition, the output pulses of 90 fs duration could be tuned from 580 nm to 610 nm. As compared with theory in both stages the fiber length chosen is larger than the optimum length (4 m,

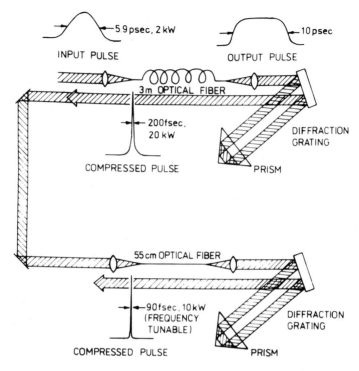

**Figure 4.9** Experimental set-up for a two-stage optical fiber compressor (from [4.37]).

3 cm) which should be one reason why the theoretical value of the possible total compression factor ($10^3$) is larger than the measured one ($\simeq 65$). Even shorter output pulses could be experimentally obtained in one stage compressors, starting, however, with the fs light pulses from a dye laser. After amplification in a series of (3–4) dye cells pumped by an actively mode locked, Q-switched Nd:YAG laser (800 Hz), a Q-switched Nd:YAG laser (10 Hz) and a copper vapor laser (5 kHz), pulses as short as 16 fs [4.46], 12 fs [4.47] and 6–8 fs [4.48], respectively were measured at the output of the fiber compressor. The corresponding experimental arrangements for the compression are similar to one of the stages sketched in Fig. 4.9 and to Fig. 4.7, respectively. Instead of gratings, prism sequences [4.33] and Gires-Tournois interferometers [4.49], [4.120] are also discussed for chirp compensation to compress the pulses behind the fiber.

The shortest pulses ever obtained experimentally have a duration of 6 fs

[4.48]⁴. However, higher orders of the dispersion than $\psi_L^{(2)}$ of the linear element had to be suitably adjusted for the spectrally broadened pulses to be compressed to nearly the transform limit. Using two different linear optical elements — a prism and a grating sequence — the quadratic and the cubic term of the phase response of this complex element could be controlled independently. (This also indicates that at pulse durations of this order, self-phase modulation and compression can be described only to a rough approximation by the nonlinear Schroedinger equation.)

*Compression after chirping of high-energy fs pulses in bulk material.* The disadvantage of the fiber compressors is that only pulses of relatively small energy can be compressed (cf. Table 4.1), where the limit is determined by the maximum possible intensity in the guided mode. As mentioned earlier, the requirement of confining the pulse intensity is for providing a certain ratio of the dispersive ($L_D$) and nonlinear ($L_{NL}$) length. Roughly speaking, longer input pulses need longer fibers. In the fs regime, however, the fiber lengths needed are in the order of several mm (cf. Table 4.1). Over such propagation distances suitably focussed Gaussian beams do not change their beam diameter very much in bulk materials. Because the beam diameter is no longer determined by the core diameter (several $\mu$m) of a single-mode fiber, but can be considerably larger, high-energy fs light pulses can be favorably chirped in this way [4.51]. Figure 4.10 shows calculated compression factors obtained with chirping in a bulk material taking into account the propagation characteristics of Gaussian beams and subsequent compression in an optimum quadratic compressor. (Alternative methods for compressing high-power pulses are described in section 4.1.6 and 4.2.4.)

*4.1.5.2 Pulse narrowing in the case of GVD and nonlinearity having different signs.* Nonlinear pulse propagation under the condition $k_L'' \tilde{\kappa}_2 < 0$ shows a completely different behavior than what was discussed in the previous section. Theoretical studies [4.52], [4.9], [4.10], [4.24] predicted the existence of pulse solutions — so-called solitons, where the pulse spreading due to the linear dispersion $k_L''$ is counterbalanced by a

----

⁴ These pulses consist of about 3 cycles of the carrier wave taken at the center of the pulse spectrum ($\lambda_L \simeq 630$ nm). It should be noted that pulses also containing only 4 oscillations were produced in the infrared ($\lambda_L = 10.4$ $\mu$m, $\tau_L = 130$ fs) [4.50] by other means, namely by cutting off the leading and trailing edges of a 100 ns $CO_2$ laser pulse. This was obtained by switching the reflection of two CdTe plates, the first to cut the leading edge and the second plate to cut the trailing edge. These reflection switches were driven by 70 fs amplified dye laser pulses and rest on refractive index changes due to the intervalence band absorption (of the 70 fs pulse). Moreover pulses of 4 optical cycles were observed in the near-infrared as a result of soliton shaping in fibers (chapters 4, 5).

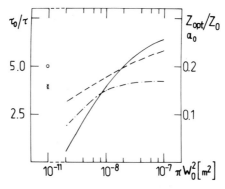

**Figure 4.10** Pulse compression of high-energy fs light pulses using bulk materials. Normalized optimum propagation length $z_{opt}/z_0$ (solid line), normalized parameter of the quadratic compressor $a_0$ (dashed line), and compression factor $\tau_0/\tau$ (dash-dotted line) as a function of the beam cross section in the beam waist located at the input of the bulk medium. For comparison, at this point the pulse intensities were assumed to be equal. The corresponding values from [4.31] referring to the use of optical single-mode fibers are indicated by "+", "×", and "o" respectively. (Parameters: $z_0 = 3.08$ cm of fused silica, $\tau_0 = 60$ fs, $P = 100$ kW, $\lambda_L = 600$ nm) (from [4.51a]).

compression mechanism resulting from the combined action of nonlinear chirp generation ($\tilde{\kappa}_2 \neq 0$) and compensation ($k_L'' \neq 0$). In this sense the two steps of pulse compression (Section 2.5) may occur simultaneously during the pulse propagation in the medium (fiber).

The first experimental observation of this phenomenon was reported in [4.53], where pulse propagation through a single mode glass fiber was studied for pulse wavelengths $\lambda_L > 1.3$ $\mu$m, which means $k_L'' < 0$, $\tilde{\kappa}_2 > 0$. These experiments confirmed the main theoretical predictions [4.53], [4.24], [4.55]. A pulse of the form

$$\overline{E}(\eta, \zeta = 0) = 1.76 \, N \sqrt{\frac{1}{|\tilde{\kappa}_2| L_D}} \, \text{sech} \left( 1.76 \frac{\eta}{\tau_{L0}} \right) \qquad (4.29a)$$

at the input of the fiber, where N is an integer, propagates without distortion for $N = 1$ (fundamental soliton). For $N > 1$ the theory yields a periodic behavior with a periodicity length

$$z_0 = \frac{\pi}{2} \frac{L_D}{(1.76)^2} \qquad (4.29b)$$

(higher-order soliton). Input pulses for which N deviates from an integer will eventually turn into a soliton. (The same holds for pulses of other shape the pulse area of which deviates from the soliton area).

The problems associated with this soliton propagation are of great interest for fiber optical communication and data transfer and have been

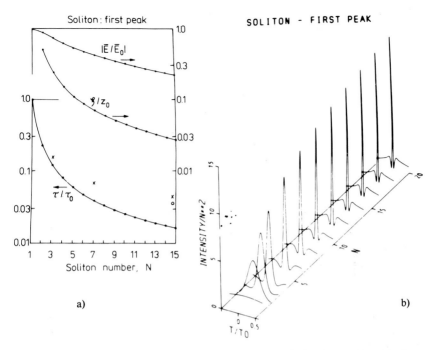

**Figure 4.11** Pulse compression due to soliton shaping, a. Calculated properties of the first optimal narrowing by means of the soliton effect in single-mode fibers and some related experimental data ($\times$, o) as a function of the soliton number, $z_0$: soliton periodicity length, b. Calculated pulse shape at the point of optimal narrowing (from [4.55]). Note, $z_0 \simeq 0.5L_D$ and $\tau_0 \simeq \tau_{L0}/1.76$.

extensively discussed theoretically as well as experimentally (e.g. [4.30], [4.54–4.64]).

With the view on pulse compression we want to point out here only that the soliton formation will always first lead to a substantial narrowing of the central pulse peak at a certain propagation length $\zeta$, no matter how complex the following behavior. The pulse splitting and narrowing was observed experimentally. For $N = 2$ the behavior is particularly simple. The pulse narrows as it propagates through the first part of the fiber achieving minimum duration at half the theoretical period. Figures 4.11a,b show some theoretical and experimental results in order to illustrate this type of pulse compression. From that the pulses become obviously shorter the higher the input intensity, i.e. the higher the order (N) of the solitons which develop. Narrowing factors up to 30 were measured. The disadvantage of this method is the relatively poor pulse quality. A considerable part of the total pulse energy is distributed over undesired broad wings and sidelobes (see Fig. 4.11b).

More detailed experimental and theoretical studies revealed that the propagation of solitons in fibers is influenced by Raman scattering. The qualitative behavior discussed refering to the first pulse narrowing, however, remains valid. For a discussion of the propagation characteristics resulting from the presence of Raman processes, see section 4.3.

It should be noted that the same nonlinear optical mechanism can also be utilized to produce a train of optical pulses in a glass fiber from an amplitude modulated cw signal coupled into the fiber [4.56]. This is a consequence of the fact that the NLSE for a cw signal exhibits an instability for a small perturbation in amplitude [4.56], [4.60] if the modulation wavelength $\lambda_m$ satisfies the relationship

$$\lambda_m > \frac{\sqrt{\pi}}{10^{4.5}\sqrt{\bar{n}}_2 A_{om}} \qquad (4.29c)$$

Recently such formations of pulse sequences were observed experimentally [4.61], where the modulation of the incident wave was produced by mixing two laser beams (Nd:YAG, InGaAsP) having different frequencies. In this way the modulation frequency which corresponds to the difference frequency determines the repetition frequency of the light pulses at the fiber output. The pulse sequence could be simply tuned by changing one of the laser frequencies (i.e. tuning of the InGaAsP laser) and repetition frequencies up to 300 GHz were reached (see Fig. 4.12).

A very high total pulse compression factor of about 1100 was reported in [4.65]. The pulses of a Nd:YAG laser at $\lambda_L = 1.319\ \mu m$ were compressed from 100 ps to 90 fs by means of a two stage compressor. The first stage consists of a usual fiber grating compressor (see Section 4.1.5.1., Table 4.1) where a 50x shortening was reached. The second stage made use of the soliton narrowing described above. In both cases optical single-mode fibers were used as nonlinear dispersive medium, where the appropriate signs of $k_L''$ ($\lessgtr$) 0 in the first (second fiber) were realized through special design of the fiber and using proper doping materials. Utilizing a similar two-stage compressor a shortening by even a factor of 2700 has been realized [4.66] leading to the generation of 33 fs pulses at 1.32 $\mu m$ (Nd:YAG laser).

*Wing reduction.* Finally we want to mention that there are methods to suppress the undesired wings which frequently accompany the pulses compressed by the fiber-grating compressor or by soliton narrowing. In the past, low power wings and subpulses were often reduced by saturable absorbers, where the time scale on which the pulse needed to be shaped determined the necessary relaxation transients of the absorber. Since most of the saturable absorbers have relaxation times well above several

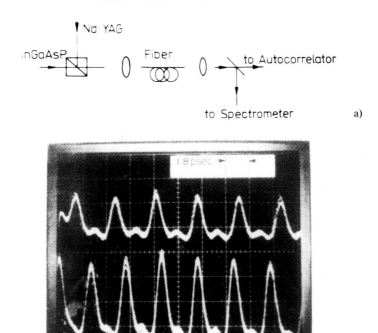

**Figure 4.12** Pulse generation through soliton shaping in fibers. The input signal is a periodically modulated cw signal obtained by mixing two laser beams which exhibit a suitable frequency spacing. a. Schematic of the experimental set-up, b. Autocorrelation traces for two different frequency detunings, resulting in different pulse repetition rates (from [4.61]).

hundred femtoseconds (with absorbing dyes, color centers, semiconductor doped glass filters, the relaxation times range from several picoseconds to nanoseconds) only the leading edge of the pulse can be shaped through saturation whereas the trailing edge remains unchanged. Of course, this causes problems if the main pulse is followed by undesired satellites. For this purpose another method was introduced in [4.67–4.69], which is based on the nonlinear birefringence effect in nonpolarization-preserving single-mode optical fibers. This effect leads to a different amount of (intensity-dependent) phase shift of the electric field components polarized along the two axes of the fiber. (These axes are determined by the polarizing direction linearly polarized low-power input light must have to produce linearly polarized output light). This induced phase shift results in a change in the output polarization. In this manner the polarization is intensity dependent at the end of the fiber. Consequently, a polarizer at the output of the fiber can block the low intensity light pulse wings and satellites, where the total transmission of the system (fiber and polarizer) was found to vary as the cube of the input intensity. Recently, this technique was combined with a simultaneous pulse compression in a fiber-grating compressor [4.70].

Other methods to clean the pulse profile rest, for example, on linear optical processes (see chapter 6).

### 4.1.6 Chirp generation through parametric interaction

In [4.71], [4.72] a technique was proposed to produce chirped pulses in parametric generators such as crystals with quadratic nonlinearity which are pumped by chirped pulses. It could be shown that under certain circumstances the chirp of the developing signal and idler pulse can be substantially larger than that of the pump pulses. Since the signal and idler pulses have nearly the same duration as the pump pulse, pulse compression through chirp compensation may yield pulses that are considerably shorter than the input pulses. As compared with the fiber-grating compressor this technique offers the advantage of the generation of compressed pulses which are tunable in wavelength and which can have higher energy.

The parametric conversion process can be understood as a three-wave interaction where the pump wave of frequency $\omega_3$ decomposes into two waves of frequencies $\omega_1$ and $\omega_2$ (e.g. [4.73], [4.74]), where

$$\omega_1 + \omega_2 = \omega_3 \tag{4.30a}$$

holds (conservation of photon energy). Besides (4.30a) the phase matching condition

$$\vec{k}_1(\omega_1) + \vec{k}_2(\omega_2) = \vec{k}_3(\omega_3) \tag{4.30b}$$

has to be satisfied for an efficient conversion process. If only the pump wave is incident $\omega_1$ and $\omega_2$ are determined by (4.30a) and (4.30b). Now we assume that the pump pulse is chirped

$$\omega_3(t) = \omega_3 + \dot{\Phi}_3(t) \tag{4.31a}$$

As the response times of suitable second-order nonlinear optical samples for parametric processes are very short, (4.30a) has to be satisfied for all t. Therefore, the output pulses are also chirped [4.71] and we have

$$\omega_1(t) = \omega_1 + \dot{\Phi}_1(t) \tag{4.31b}$$

$$\omega_2(t) = \omega_2 + \dot{\Phi}_2(t) \tag{4.31c}$$

where

$$\dot{\Phi}_1(t) + \dot{\Phi}_2(t) = \dot{\Phi}_3(t) \tag{4.31d}$$

and

$$\vec{k}_1(\omega_1(t)) + \vec{k}_2(\omega_2(t)) = \vec{k}_3(\omega_3(t)) \qquad (4.31e)$$

For linearly chirped pulses ($\ddot{\Phi}_i(t)$ = const.) to which we restrict our consideration, and provided that $|\dot{\Phi}_i| \ll \omega_i$ equation (4.31e) can be expanded into a Taylor series and yields

$$\frac{\dot{\Phi}_1}{u_1} + \frac{\dot{\Phi}_2}{u_2} = \frac{\dot{\Phi}_3}{u_3} \qquad (4.32)$$

where $u_i = (d\omega/dk)^{-1}|_{\omega_i}$ is the phase velocity at frequency $\omega_i$. With (4.32) and (4.31d) the relationship between the chirp parameters of the pump and the output pulses can be written as

$$\dot{\Phi}_1 = r_0 \dot{\Phi}_3 \qquad (4.33a)$$

$$\dot{\Phi}_2 = (1 - r_0)\dot{\Phi}_3 \qquad (4.33b)$$

where

$$r_0 = \frac{u_3^{-1} - u_2^{-1}}{u_1^{-1} - u_2^{-1}} \qquad (4.33c)$$

Obviously the phase modulation of the signal and idler pulse can exceed that of the pump pulse considerably, if the group velocities $u_1$ and $u_2$ differ only slightly from each other. However, for the relation between $\dot{\Phi}_{1,2}$ and $\dot{\Phi}_3$ to remain linear,

$$|\omega_1 - \omega_2| \gg |\ddot{\Phi}_{1,2} \tau_{L1,2}| \qquad (4.34)$$

must be required. Otherwise terms of higher order would have to be considered in the expansion of (4.31e) with the consequence that (4.33a,b) become nonlinear equations. As outlined in section 3, complicated nonlinear phase modifications are not desirable with the aim to compress the pulses. In this way (4.34) in connection with (4.33a,b) determines an upper limit for the chirp as well as the lowest value the difference $\omega_1 - \omega_2$ can take on. Using

$$\frac{1}{u_1} - \frac{1}{u_2} = \frac{dk}{d\omega}\bigg|_{\omega_1} - \frac{dk}{d\omega}\bigg|_{\omega_2} \simeq \frac{d^2k}{d\omega^2}\bigg|_{\omega_1}(\omega_1 - \omega_2) \qquad (4.35)$$

we can write (4.33a) as

$$\ddot{\Phi}_1 = \frac{u_3^{-1} - u_2^{-1}}{\omega_3 |d^2k/d\omega^2|_{\omega_1}}\left(1 - 2\frac{\omega_2}{\omega_3}\right)^{-1}\ddot{\Phi}_3 \qquad (4.36)$$

Figure 4.13 shows the chirp enhancement $\ddot{\Phi}_1/\ddot{\Phi}_3$ and $\ddot{\Phi}_2/\ddot{\Phi}_3$ as a function of

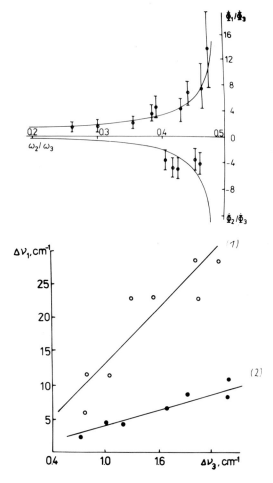

**Figure 4.13** Chirp production through parametric generation, a. Chirp enhancement $\dot{\Phi}_i/\dot{\Phi}_3$ through parametic interaction as a function of $\omega_2/\omega_3$ (I: experimental data), b. Frequency change of the signal pulse $\Delta\nu_1 = \dot{\Phi}_1\tau_1/c$ as a function of the frequency change of the pump pulse $\Delta\nu_3$ for $\omega_2/\omega_3 = 0.428$ (2), (from [4.71]).

$\omega_2/\omega_3$ according to (4.31d) and (4.36) where the material parameters of LiNbO$_3$ crystals were used.

The experimental data [4.71] were gained by pumping the crystal with the second harmonic of a Nd:YAG laser ($\tau_{L3} \simeq$ 30 ps, $\lambda_3 =$ 530 nm, $\ddot{\Phi}_3\tau_{L3} \simeq 5.4 \times 10^{10}$ s$^{-1}$). The phase modulation of these pump pulses is a result of the Kerr nonlinearity in the laser medium.

Theoretical estimations show that starting with these pulses a chirp enhancement up to $\ddot{\Phi}_1/\ddot{\Phi}_3 \simeq 50$ should be possible, which would result in $\tau_{L1} \simeq 150$ fs after pulse compression through chirp compensation. In this manner it was possible to generate 280 fs pulses through parametric amplification, chirp enhancement and subsequent compression in a grating compressor [4.72b].

While the generation of chirped ps pulses in optical fibers is restricted to pulse energies in the nJ range due to the strong focussing (higher non-linearities, Raman effect, damage threshold) the parametric oscillation process can be advantageously used also for pulse energies in the mJ range. Dealing with conversion bandwidths corresponding to femto-second light pulses the disadvantage of these methods originates from the effect of GVD which comes into play if the crystal length is on the order of the dispersion length $L'_D$, (see (4.12)). The GVD influences the conversion efficiency and it disturbs the linear relationships between $\dot{\Phi}_3$ and $\dot{\Phi}_{1,2}$.

It is interesting to note that the initial chirp can be impressed on the pump pulse by linear optical processes, too, which offers rather simple ways of phase shaping. As an example we consider an unchirped Gaussian pump pulse (duration $\tau_{L0}$) which is first sent through a dispersive sample exhibiting GVD (e.g., glass of length L). According to (3.3) the pulse duration increases due to dispersion and reaches at L

$$\tau_L(L) = \tau_{L0} \sqrt{1 + \overline{A}} \qquad (4.38a)$$

where

$$\overline{A} = (4 \ln 2)^2 \left( \frac{k''_L L}{\tau_{L0}^2} \right)^2 \qquad (4.38b)$$

Simultaneously a linear chirp develops ($\Phi(t) = 2\beta t^2$) where the chirp parameter is given by

$$\beta(L) = \frac{\sqrt{A} \ln 2}{\tau_{L0}^2} \qquad (4.39)$$

(see equation (3.3d)). This pulse serves now as the pump pulse in the parametric process, where the chirp is increased by a factor $r_0$ (for pulse 1). The pulse duration remains unaffected in the ideal case to be discussed (i.e. without considering GVD in the crystal). Thus, we obtain for the chirp parameter

$$\overline{\beta} = r_0 \beta(L) \qquad (4.40)$$

The chirped output pulse can be compressed in another linear element exhibiting GVD of the appropriate sign, that is of opposite (equal) sign for pulse 1 (2) as compared with the first linear element. The maximum compression (see section 3) at $L_\beta = L_\beta(\beta)$ produces unchirped Gaussian pulses with a duration

$$\tau_{Lc} = \tau_L(L) \left( 1 - \frac{r_0^2 \overline{A}}{4 + r_0^2 \overline{A}} \right) \qquad (4.41)$$

The total compression factor R is gained from (4.38a), (4.41) and is

$$R = \frac{\tau_{L0}}{\tau_{Lc}} = \left[ (1 + \overline{A}) \left( 1 - \frac{r_0^2 \overline{A}}{4 + r_0^2 \overline{A}} \right) \right]^{-1/2} \qquad (4.42)$$

For $r_0^2 \overline{A} \gg 1$ we find

$$R \simeq \frac{r_0}{2} \sqrt{\frac{\overline{A}}{1 + \overline{A}}} \qquad (4.43)$$

Consequently, with the initial pulse chirping due to GVD, pulse compression factors on the order of $r_0/2 \simeq 20$ seem to be possible. If we assume a finite value of $\overline{A} = 3$ corresponding to a glass path that doubles the initial pulse duration we find a possible compression of $R \simeq 17$ for $r_0 = 40$.

## 4.2  Self-phase modulation through near-resonant nonlinear interaction

Pulses propagating through a saturable quasiresonant[5] medium are changed not only in their modulus but also in their phase, which as in linear optics, is a result of the fact that a certain dispersion profile corresponds to each resonance line (section 3). The striking difference compared to the linear optical case is a temporal change of the dispersion curve caused by saturation of the resonant transition. This self-induced refractive index change leads then to self-phase modulation associated with spectral broadening.

Initially, this type of phase modulation was utilized in experiments on optically bistable resonators [4.75], [4.3] and it later gained importance in the pulse formation in femtosecond lasers [4.76–4.86] as well as in connection with pulse amplification (see section 4.2.4).

### 4.2.1  Basic equations

For the discussion of the main features of self-phase modulation due to quasiresonant interaction, we will make some restricting assumptions.

First we restrict outself to a homogeneously broadened two-level system

---

[5] Quasiresonant medium means that for its transition frequency $\omega_{ij}$ and spectral width $\Delta\omega_{ij}$ $|(\omega_{ij} - \omega_L)/\Delta\omega_{ij}| \leqslant 1$ holds and that the influence of further resonance lines can be neglected.

which in particular implies a coincidence of the absorption profile and the profile for stimulated emission during the interaction.

Second, the pulses are to be longer than the phase relaxation time $T_2$ of the transition and the pulse area is assumed to satisfy the relation

$$\left| \frac{\mu_{12}}{\hbar} \right| \int\limits_{-\infty}^{\infty} |\overline{E}(t')| \, dt' \ll 1 \qquad (4.44)$$

($\mu_{12}$: transition moment).

Therefore, we will not consider effects associated with coherent pulse propagation (an extensive treatment can be found, e.g., in [4.87]).

It should be noted that these assumptions are valid only in approximation when considering the interaction of fs light pulses with organic dye molecules. Recent experimental results in fs time-resolved spectroscopy that require a more complex model for their interpretation are:

(1) the observation of spectral hole-burning in the absorption profile indicating inhomogeneous contributions to the line broadening [4.48]
(2) the measurements of quantum beats exhibiting relatively long dephasing times [4.89], [4.90]
(3) the difference in the rate parameters measured for the absorption recovery [4.91] and the rise of the fluorescence [4.92]

At the end of section 4.2.2 we will briefly discuss the modifications of the self-phase modulation in the presence of inhomogeneous broadening and in the case that relaxation processes occur between vibrational levels in the excited electronic state during the pulse. On a ps time scale the dye molecules can be regarded as an ensemble of homogeneously broadened four-level systems having different absorption and fluorescence profiles. Accordingly the resulting dispersion profile is obtained from the (weighted) sum of these two profiles [4.93].

For our discussion here the starting point is again the wave equation (2.12) and the polarization (2.13). The linear part of the polarization $P^L$ is attributed to the host medium (e.g. solvent) of the resonant particles which in turn are described by the nonlinear polarization $P^{NL}$. For $P^L$ the same discussion holds as in the derivation of (2.20a). Applying the SVEA we will again consider only the first term on the right hand side of (4.2). In order to describe pulse propagation under conditions to be discussed here, we only have to specify the nonlinear polarization which in terms of the density matrix is given by the off-diagonal element (e.g. [4.87])

$$P^{NL}(t,z) = \tfrac{1}{2} \, \overline{P}^{NL}(t,z) e^{i(\omega_L t - k_L z)} + \text{c.c.}$$
$$= q\mu_{21} \, S_{12}(t,z) + q\mu_{12} \, S_{21}(t,z) \qquad (4.45)$$

(q: particle number density of the atomic systems, $S_{ij}$: off-diagonal element of the density matrix). $S_{12}(t,z)$ can be determined from the density matrix equations describing the interaction of the atomic system and the electric field

$$\left(\frac{\partial}{\partial t} - i\omega_{12} + \frac{1}{T_2}\right) S_{12}(t) = -\frac{i}{2\hbar} S_1(t)\mu_{12} \,\overline{E}(t,z)e^{i(\omega_L t - k_L z)} \qquad (4.46)$$

$$\left(\frac{\partial}{\partial t} + \frac{1}{T_1}\right) S_1(t) = -\frac{2i}{\hbar} (\mu_{21} S_{12}(t) - \mu_{12} S_{21}(t)) \,\overline{E}(t,z) \qquad (4.47)$$

($S_1 = 2S_{11} - 1$, $S_{11}$: diagonal element, $T_1$: energy relaxation time, $T_2$: phase relaxation time, $\omega_{21}$: resonance frequency of the transition). In the frequency domain $S_{12}$ reads [4.94]

$$\underline{S}_{12}(\omega) = -\frac{i}{2\hbar} \mu_{12} T_2 \, e^{i(\omega_L t - k_L z)} \mathscr{F} \times$$

$$\times \mathscr{F}\left\{ S_1(t) \,\overline{E}(t,z) \sum_{m=0}^{\infty} [-i\mathscr{L}T_2(\omega - \omega_L)]^m \right\} \qquad (4.48a)$$

where $\mathscr{F}$ denotes the Fourier transform and

$$\mathscr{L} = L_r + iL_i = [1 + iT_2(\omega_L - \omega_{21})]^{-1} \qquad (4.48b)$$

is the complex line shape function with $L_i$ and $L_r$ as real and imaginary part, respectively.

An inspection of (4.48a) shows that, considering only the term with $m = 0$, the characteristics of the transition are taken at a fixed value $\omega = \omega_L$ i.e. their change throughout the spectrum of the interacting pulse is neglected. This case ($m = 0$) leads to the so-called rate equation approximation (REA) which according to (4.48a) requires $|T_2(\omega - \omega_L)| \ll 1$ or the equivalent relation

$$\Delta\omega T_2 \ll 1 \qquad (4.49)$$

and which describes pulse shaping as a result of saturation only.

In order to discuss some aspects of the interplay between the linear (dispersive) and nonlinear (saturation) effects, we will account for terms up to $m = 2$. Thus, we obtain in terms of local coordinates (section 2) from (4.2), (4.45), (4.47) and the backtransformed equation (4.48a)

$$\left(\frac{\partial}{\partial \zeta} - D_1\right) \overline{E}(\eta, \zeta)$$

$$= -\frac{1}{2} \sigma_{12}\mathscr{L} \left(1 - \mathscr{L}T_2 \frac{\partial}{\partial \eta} + \mathscr{L}^2 T_2^2 \frac{\partial^2}{\partial \eta^2}\right) q_1(\eta, \zeta) \,\overline{E}(\eta, \zeta) \quad (4.50a)$$

for the complex pulse envelope, and

$$\frac{\partial}{\partial\eta} q_1(\eta,\zeta) = -2\beta L_r q_1(\eta,\zeta) \, |\bar{E}(\eta,\zeta)|^2 \tag{4.50b}$$

$$+ 2\beta \, \mathrm{Re}\left\{ \mathscr{L}\bar{E}^*(\eta,\zeta)\left[ \mathscr{L}T_2 \frac{\partial}{\partial\eta} - \mathscr{L}^2 T_2^2 \frac{\partial^2}{\partial\eta^2} \right] q_1(\eta,\zeta)\bar{E}(\eta,\zeta) - \frac{q_1(\eta,\zeta) - q_0}{T_1} \right.$$

for the occupation inversion $q_1 = qS_1$. Here

$$\sigma_{12} = \mu_0 |\mu_{12}|^2 \, T_2 \, \omega_L / (\hbar k_L) \tag{4.50c}$$

is the interaction cross section in the line center ($\omega_L = \omega_{21}$) and

$$\beta = \tfrac{1}{2} T_2 |\mu_{12}|^2 / \hbar^2 \tag{4.50d}$$

Note that for a maximum amplifying (absorbing) medium $S_1 = -1(1)$ holds.

Roughly speaking, the term with the second derivative on the right side of equation (4.50a) is responsible for the dispersion caused by the finite width of the transition.

## 4.2.2 Neglect of the dispersion of the host medium

The dispersion of the host medium can be neglected ($D_1 = 0$ in (4.50a)) if for the sample length L

$$L \ll L_D \tag{4.51}$$

holds, which according to (3.4) means that the pulse distortion due to GVD alone is negligible. Considering femtosecond light pulses this condition is fulfilled in good approximation, for example, in typical passively modelocked dye lasers, where the interaction lengths (in the saturable media) are on the order of 0.1 mm.

For pulses much longer than the phase decay time $T_2$ the time derivatives in the right sides of (4.50a,b) can be omitted (REA) and the simplified system of equations can be solved analytically for the two limiting cases in which the pulse duration $\tau_L$ is much shorter (longer) than the energy relaxation time $T_1$ of the transition. It should be noted that with fs light pulses the case of energy saturation ($\tau_L \ll T_1$) mostly occurs.

The equations for modulus $A(\eta)$ and phase $\Phi(\eta)$ can be solved separately and we find for the instantaneous frequency shift using the solution for the modulus [4.95]

$$\frac{d\Phi}{d\eta} = \frac{1}{2} \frac{L_i}{L_r} \frac{e^{\kappa L_r} - 1}{e^{\kappa L_r} - 1 + e^{\bar{e}(\eta)}} \frac{d\bar{e}(\eta)}{d\eta} \tag{4.52a}$$

for the case of energy saturation ($\tau_L \ll T_1$) and

$$\frac{d\Phi}{d\eta} = \frac{1}{2} \kappa L_i \frac{1}{(1 + T_1 F(\eta))^2} \frac{dF(\eta)}{d\eta} \qquad (4.52b)$$

if intensity saturation ($T_1 \ll \tau_L$) is present [4.96]. In (4.53b) we have assumed for simplicity that the absolute value of

$$\kappa = \sigma_{12} L\, q_1(\eta \rightarrow -\infty) \qquad (4.53a)$$

is very small compared to 1. $F(\eta) = \beta L_r |\overline{E}_0|^2$ is the photon flux density at the input of the sample, normalized to the interaction cross section at the laser frequency $\sigma_{21} L_r$, and the pulse energy $\bar{\epsilon}(\eta)$ normalized to the saturation energy is given by

$$\bar{\epsilon}(\eta) = \int_{-\infty}^{\eta} d\eta'\, F(\eta') \qquad (4.53b)$$

where $\bar{\epsilon}(\infty)$ is the normalized total pulse energy.

In both cases, the phase modulation results from the temporal change of the dispersion curve caused by saturation. Figure 4.14 shows the temporal behavior of the instantaneous frequency after transit through a weak

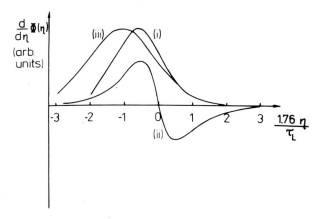

**Figure 4.14** Instantaneous frequency shift versus local time after transit through an absorber with energy saturation (curve (i)), an absorber with intensity saturation (curve (ii)) and strong amplifier with weak energy saturation (iii). The laser frequency was chosen to be $\omega_L < \omega_{21}$ for the absorber and $\omega_L > \omega_{21}$ for the amplifier.

absorber saturated by energy (i), by intensity (ii) and a strong amplifier with weak energy saturation (iii).

A closer inspection of (4.52a,b) shows that the sign of the phase modulation is determined by the sign of $\kappa L_i$, i.e., the relative position of the pulse mid frequency $\omega_L$ and the center frequency of the transition $\omega_{21}$ and whether an absorbing ($\kappa > 0$) or amplifying ($\kappa < 0$) interaction takes place. The magnitude of the phase modulation taken throughout the half width of the pulse is related to $|(\omega_L - \omega_{21})T_2|$ at given $\kappa$ and reaches a maximum at certain values of the normalized intensity $T_1F$ and energy $\bar{\epsilon}(\infty)$, respectively, for a given shape of the input pulse [4.83]. The latter is a consequence of the fact that with high energies and intensities, respectively the phase variation and with it, the frequency sweep takes place only at the very leading part of the pulse. Accordingly, the main part of the pulse exhibits almost no phase modulation.

Although the frequency changes nonlinearly with time (Fig. 4.14), one can find a monotonic, almost linear behavior around the pulse center indicating the possibility of pulse compression with suitable dispersive elements. This is illustrated in Fig. 4.15, which shows the phase modulation and pulse duration of a pulse initially chirped (and shaped) by a weak saturable absorber at the output of a linear element with variable GVD (quadratic compressor). Obviously, the pulse duration attains a minimum at a certain amount of GVD where the phase modulation in the pulse

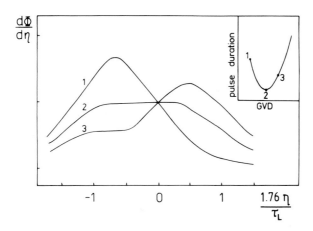

**Figure 4.15** Temporal behavior of the instantaneous frequency and pulse duration for different values of the GVD. The $\text{sech}^2(1.76\,\eta/\tau_L)$ pulse was initially shaped and chirped by an absorber, whose saturation was controlled by the energy. (from [4.96])

center is almost compensated. Note that the compression due to chirp compensation only is shown, and with strong saturation substantially larger effects can be achieved as a result of pulse steepening and wing reduction caused by the time dependent transmission which is not to be discussed here (see, e.g. [4.97]). Because the achievable compression is rather small (0.1 . . . 2% [4.96]) these effects can hardly be utilized for efficient extracavity pulse compression but come into play in intracavity processes as in the pulse formation in lasers due to the multiple passage through such elements.

So far our discussion has been within the frame of rate equations which can no longer be used if the pulse duration approaches the phase relaxation time $T_2$ of the transition. A general treatment requires extensive numerical calculations to solve (4.46), (4.47), (4.50a) [4.98]. Here, we only want to discuss the effects of a finite phase memory by means of approximate solutions in the limit of weak saturation ($\exp(-\bar{\epsilon}) \simeq 1 - \bar{\epsilon} + \frac{1}{2}\bar{\epsilon}^2$) and provided that $\tau_L \geqslant T_2$. Starting with (4.50a,b) we can apply an iterative procedure for evaluating the pulse envelope behind the sample and find [4.99]

$$
\begin{aligned}
\bar{E}(\eta,L) = \Bigg\{ 1 &- \frac{1}{2}\,\kappa\mathscr{L}\left[ 1 - \bar{\epsilon}(\eta) + \frac{1}{2}\bar{\epsilon}^2(\eta) + T_2\mathscr{L}F(\eta) \right.\\
&\left. + 2\beta T_2 \int_{-\infty}^{\eta} d\eta'\, \mathrm{Re}\,\left( \mathscr{L}^2\bar{E}_0^*\,\frac{d}{d\eta'}\bar{E}_0 \right) \right]\\
&+ \frac{1}{2}\,\kappa\mathscr{L}^2\,T_2(1-\bar{\epsilon}(\eta))\,\frac{d}{d\eta} - \frac{1}{2}\,\kappa T_2^2\mathscr{L}^3\,\frac{d^2}{d\eta^2} \Bigg\}\,\bar{E}_0(\eta) \quad (4.54)
\end{aligned}
$$

For $T_2 \to 0$ we obtain the corresponding expression in the REA for weak saturation. The interplay between saturation and the dispersive properties of the transition is to be discussed here only with respect to its influence on the pulse duration (for a more detailed discussion see [4.100], [4.93b]). For this we consider the chirp and the pulse duration as a function of the detuning from $\omega_{21}$ at fixed values of $\kappa L_r$ and $\bar{\epsilon}(\infty)$, see Fig. 4.16. Obviously, there are regions for the detuning where the combined action of nonlinear and linear processes lengthens or shortens the pulses as compared with the pulses shaped by saturation only (REA). This shortening can also be interpreted as a result of chirping and subsequent compression taking place simultaneously in the quasiresonant medium.

As mentioned earlier the description of the quasiresonant transition

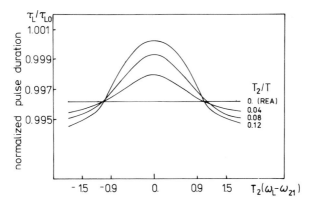

**Figure 4.16** Normalized pulse duration after passing a $\mathrm{sech}^2(1.76\,\eta/\tau_L)$ pulse through a weak saturable absorber as a function of the detuning from the center frequency for different values of the phase relaxation time $T_2$. ($\kappa L_R$ was kept constant) (from [4.100]).

with homogeneously broadened two-level systems might be too rough an approximation. Therefore, chirping under more general conditions was investigated for a four-level system with the aim to study the influence of a finite vibronic relaxation during the pulse interaction [4.93b]. Furthermore, the inhomogeneous contributions to the line broadening were discussed [4.93c]. In both cases the main conclusions drawn before remain valid, though, in view of pulse compression there exist parameter ranges where the magnitude of a favorable linear frequency sweep across the pulse center can be larger.

### 4.2.3 Combined action of the dispersion of the host medium and quasiresonant nonlinearity

As discussed above $D_1 \neq 0$ has to be considered if the sample length approaches $L_D$. Several years ago, this situation was discussed in conjunction with the steady-state pulse propagation through long (infinite) amplifying media ($\pi$-pulse) in the presence of host dispersion [4.101]. Later the interplay between self-phase modulation (Kerr type) and dispersion resulting from resonant transitions was studied [4.102] under conditions where the saturation could be neglected.

Dealing with femtosecond light pulses, the simultaneous presence of GVD and quasiresonant nonlinearity (saturation) already has to be considered for example in pulse amplifiers built up with dye cells of several cm in length. In this case, the amplified pulse is no longer determined by the equations derived in the REA [4.95], but the GVD (or, if necessary, higher-order dispersion terms) has to be considered in the wave equation.

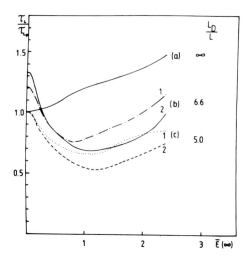

**Figure 4.17** Pulse duration of an amplified sech pulse as a function of the saturation calculated within the frame of the REA. a. neglect of GVD ($L_D/L = \infty$), b. under the presence of GVD, c. pulses from b, after an ideal compressor. Parameters: small signal amplification $10^5$, $(\omega_L - \omega_{21})T_2 \simeq -1.25$ (from [4.103]).

An analytical evaluation of the resulting system of differential equations, however, is not possible. Figure 4.17 summarizes numerical results [4.103] describing the pulse duration after amplification with and without influence of GVD for two different lengths of the cell as compared with the dispersive length. The pulse duration is depicted as a function of the saturation which was assumed to be constant along the cell for simplicity. In experiments the latter would require a proper defocussing. Obviously the pulse spreading due to the amplification process can be counter-balanced by the simultaneous chirp production and pulse compression. Moreover, the net phase modulation at the output of the amplifier provides the possibility of further compression (here shown when using an ideal linear element).

### 4.2.4  Experimental results

Of course, due to the smallness of the effects calculated in 4.2.2 they can hardly be measured extracavity and separately from other processes. In section 5 we will discuss experimental results of intracavity chirping, where they play a significant part.

The refractive index change in dye solutions induced by saturation through ps excitation was measured in [4.93a]. On this time scale the dye molecules can be represented as 4-level systems where absorption and emission have different transition profiles. Accordingly, the experimental findings could be explained with an effective dispersion profile containing the dispersion resulting from the absorbing two-level system as well as from the amplifying two-level system.

Larger effects could be measured dealing with semiconductors [104] even at relatively small saturation. Due to their strong nonlinearity these materials are also very attractive for PPM where the wavelength of the signal pulse can be in the transparent region of the sample.

In amplifier chains for femtosecond light pulses group velocity dispersion and self-phase modulation proceed simultaneously and result in chirped output pulses. Thus, most of the fs pulse amplifiers are terminated by a compression stage (grating pair, prism sequence, Gires–Tournois interferometer). For example, the first amplifier for fs light pulses [4.105a] based on a frequency-doubled Nd:YAG laser used a grating pair to compress the amplified pulses to almost their original duration of 65 fs. A similar recompression to 40 fs could be achieved in a high repetition rate amplifier pumped by a Copper — vapor laser [4.105b]. The recompression to the initial pulse duration indicates that the pulse broadening due to GVD and the saturable amplification itself is compensated, which means that nonlinear optical processes leading to chirp generation may play a part.

Figure 4.18 shows an excimer laser pumped fs pulse amplifier where the fourth stage is followed by a prism pair made of highly dispersive optical glass (SF2). The 100 fs input pulses from a CPM laser (616 nm) were amplified and broadened to 230 fs which was accompanied with a spectral broadening. After the passage through the prism compressor, a pulse

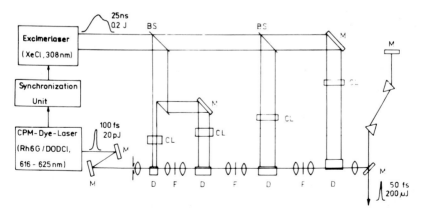

**Figure 4.18** Combined femtosecond pulse amplification, self-phase modulation and compression (from [4.106]).

duration of 55 fs was measured, which means a total shortening of about 2 as compared with the pulses from the oscillator. Possible reasons for this favorable pulse shaping and spectral broadening are fast refractive index changes in the color glass filters used as saturable absorbers and the combined action of dispersion, amplification and chirp generation in the dye cells which, however, are connected to each other in a complicated manner. Recently a direct measurement of the chirp produced by saturable amplification of ps pulses was reported in [4.107].

There are also concepts to chirp and lengthen light pulses in an optical fiber before they are amplified [4.108], [4.109], [4.34]. In particular when dealing with solid state laser pulses where the pulse intensity is critical in the amplification stages because of the occurrence of undesired nonlinear effects, the pulse broadening in the fibers can be advantageously used to keep the pulse intensity small enough while the energy amplification can be relatively high. In this manner pulses from a cw Nd:YAG laser (100 ps, 3 nJ) could be amplified and compressed to 100 mJ and 2 ps [4.108]. Limits of this compression technique is the finite width of the gain profile which has to be able to amplify the pulse spectrum broadened by the fiber.

## 4.3   Pulse propagation with nonresonant nonlinearity (Kerr effect), resonant nonlinearity (stimulated Raman scattering, SRS) and linear dispersion

At the end of section 4.2.3 we briefly discussed the combined action of resonant and nonresonant nonlinearity as well as dispersion in light pulse propagation for a dispersive amplifier. Here we would like to present nonlinear pulse propagation through optical fibers influenced by Raman gain, which has recently gained much attention, both in experimental [4.110–4.124] and theoretical [4.125–4.132] work.

In the treatment of the pulse propagation through optical fibers in section 4.1 we neglected the occurrence of other nonlinear optical processes. Such effects might influence the processes of chirp production, pulse compression and soliton formation considered here, as they affect the intensity and the temporal phase of the pump pulses. Furthermore, secondary light pulses at other carrier frequencies might be generated by such nonlinear effects, and these pulses can also be chirped or be soliton-like.

The SRS is of particular importance in this connection, as it is of the same nonlinear order as the self-phase modulation dealt with here and the nonlinear susceptibility of the Raman process is larger than that of the Kerr effect for certain conditions. In optical glass fibers (as well as in bulk materials) the broad Raman gain profile for the Stokes pulse extends up to

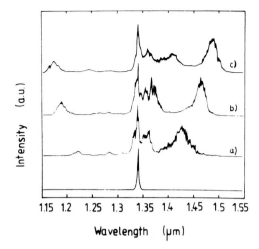

**Figure 4.19** Spectra of the pulse after three different lengths of fiber (a) 17 m, (b) 50 m, (c) 150 m. For comparison the input spectrum is also shown. Obviously a strong Stokes and a weak anti-Stokes band emerge, where the long wavelength-part of the pump pulse spectrum broadened by SPM is depleted. Input pulse parameters: $\lambda_L = 1.341$ $\mu$m, $\tau_L = 0.83$ ps, $I_L = 0.66$ GW/cm. (From [4.124]).

the frequency of the pump pulse [4.132], where because of the extended spectrum of the pump pulse, there is a overlapping region. Therefore, the lower-frequency components of the pulse experience Raman gain which in turn is pumped by the higher-frequency components. Furthermore, Raman pulses originate from the amplification of spontaneously scattered light. This, finally, leads to the development of a Stokes pulse which separates from the pump pulse after the walk-off distance due to GVD. Depending on the sign of GVD, i.e. on the position of the pulse wavelength with respect to the zero-dispersion wavelength, these processes proceed simultaneously with SPM or with soliton shaping. For illustration Fig. 4.19 shows the development of the pump and Raman pulse spectrum as the pulse propagates through the fiber.

With long input pulses ($\tau_L > 100$ fs) the threshold of SRS in single-mode fibers is rather low [4.134], [4.112], [4.133]. The influence of Raman scattering decreases with decreasing pulse width. The noninstantaneous response of SRS (memory effect) is determined by the shape of the Raman spectrum, the spectral width of which is about 250 cm$^{-1}$ and the maximum is shifted from the pump wavelength by about $\Delta\tilde{\nu} = 440$ cm$^{-1}$.

In the simplest case, i.e. neglect of dispersion and assuming memory-free response, the threshold power $P_{th}$ for SRS is given by [4.135]

$$P_{th} \simeq 30 \frac{A_{eff}}{gL_F} \qquad (4.55)$$

where $A_{eff}$ is the effective core area of the fiber, g is the Raman gain ($\simeq 9.2$ $10^{-12}$ cm/W at 1.06 $\mu$m in fused silica [4.134]) and $L_F$ is the length of the fiber. For $A_{eff} \simeq 3 \ 10^{-7}$ cm$^2$ and $L_F = 50$ m a threshold power of $P_{th} \simeq 200$ W is expected.

In the following, we want to review important experimental results of nonlinear pulse propagation through optical fibers under the influence of SRS, discuss the influence of SRS on pulse compression and soliton shaping as well as briefly describe several theoretical models which start from a modified NLSE and which can explain most of the experimental findings.

### 4.3.1   Influence of SRS on pulse compression

In section 4.1 we pointed out that the pulse propagation in the region where $k_L'' \bar{n}_2 > 0$ leads to almost rectangular pulses. These pulses exhibit a linear dependence of the frequency with time which enables a very effective pulse compression in a subsequent linear element with GVD. Above SRS-threshold this model does not hold any longer, because the generation of Stokes pulses changes the properties of the pump pulse considerably. The generation of the latter is in general a rather complex process as it is affected by the pulse walk-off, the noninstantaneous response of the material and the depletion of the laser pulse. The walk-off length $l_w$ (i.e. the propagation length after which pump and Stokes pulse are separated by the initial pulse duration $\tau_L$) is given by

$$l_w = \frac{v_S v_L}{v_S - v_L} \tau_L \qquad (4.56)$$

($v_S$, $v_L$: group velocity of Stokes and pump pulse, respectively).

First we want to consider the case when GVD can be neglected, i.e., when broadening and separation of pump and Stokes pulse do not occur. Experimentally this situation is met for $L_F < L_D$, $l_w$ and sufficiently high pulse intensities. Under these conditions the combined action of SPM and SRS gives rise to pulse fragmentation which was experimentally proved with ns [4.136] and ps light pulses [4.111], [4.120] and can be explained as follows. Since the SRS is proportional to the pump pulse intensity, the center of the pulse is preferentially converted into Stokes light until complete depletion. The remaining parts of the pulse corresponding to the wings of the initial pump pulse form the fragments. In the same manner the first Stokes decomposes into two fragments due to the pumping into the second Stokes and so on. Figure 4.20 shows autocorrelation traces of

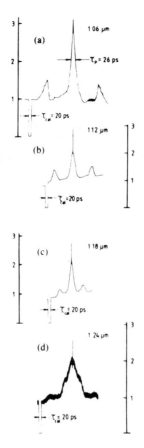

**Figure 4.20** Autocorrelation traces of the (a) fundamental at 1.06 μm, (b) first Stokes at 1.12 μm, (c) second Stokes at 1.18 μm and (d) third Stokes pulse at 1.24 μm showing pulse fragmentation and depletion. ($L_F = 4$ m, fiber core diameter $d = 7$ μm, $\lambda_L = 1.06$ μm, $P_L = 100$ kW, $\tau_L = 90$ ps) (from [4.120]).

the pump and several Stokes pulses, where the satellites clearly indicate the presence of fragments. In [4.120] the authors also investigated the possibility of compressing these fragments. It was found that the fragments have a negative chirp, which probably results from the negative chirp of the wings of the pump pulse due to SPM (cf. Figure 4.2). The fragments of the fundamental could be compressed by a factor 5 using a Gires-Tournois interferometer assembly.

Next we consider the propagation in fibers long compared to $l_w$ or in the order of $l_w$. If the fiber is longer than several walk-off lengths, i.e. in the

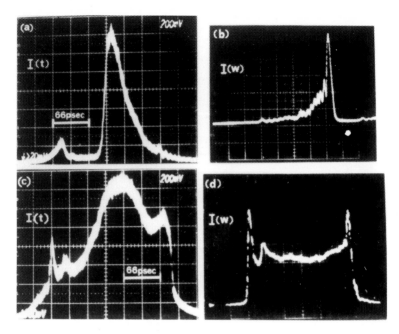

**Figure 4.21** Temporal and spectral pulse shape after 37 m ((a), (b)) and 1 km ((c), (d)) fiber. Experiments were performed with 100 ps Nd:YAG laser pulses at 1.06 μm and mean power (a,b) P = 3.3W (Raman threshold at 2.2W) (c,d) P = 0,74W (Raman threshold at 0.52W) (from [4.116]).

case when GVD plays a part, the Stokes pulse separates from the pump pulse. It could be shown that the maximum of the Raman signal is produced about 1.5–4 walk-off distances into the fiber [4.118], [4.122], but can be considerably lesser or greater depending on the pump power and Raman conversion, respectively. Because of the walk-off the Raman interaction length is strongly limited. With normal GVD ($v_S > v_L$) the Stokes pulse gradually advances the pump pulse. Walking through the pump it depletes preferentially the red shifted part of the pump spectrum which can be found at the leading edge of the pulse. As a consequence, the Stokes pulse will arrive earlier at the exit of the fiber.

The question arises about the compressibility of the depleted pump pulse and the Stokes pulse under these conditions. The properties and the compression of the pump pulses were investigated in [4.116], [4.121]. Figure 4.21 shows the temporal and spectral pulse shapes after two different fiber lengths (37 m, 1 km). In the short fiber SRS depletes the pulse asymmetrically, which leads to strong reduction of spectral width and spectral asymmetry. Therefore it is recommended to work below SRS

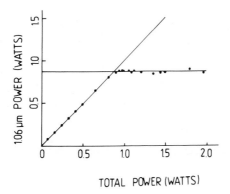

**Figure 4.22** Power of the depleted pump pulse at the output of the fiber versus the power of the input pulse. ($L_F = 145$ m, $\tau_L = 75$ ps, repetition rate 100 MHz) (from [4.121]).

threshold with short fibers. In the long fiber the combined effect of SPM and GVD again broadens the pulse spectrum and reshapes the pulse after the walk-off of the Stokes pulse. Here SRS reduces only the pulse energy, whereas the spectrum evolves symmetrically up to a bandwidth four times larger (4.5 nm) as compared with the short fiber [4.116]. Under optimum conditions the compression factor was about 100 (100 ps to 1 ps) and was obtained in one stage working with Nd:YAG laser pulses at 1.06 $\mu$m [4.116].

In [4.121] the interesting fact was found that due to SRS the pump pulse energy can be stabilized working above a critical input power, see Fig. 4.22, which is a consequence of power dependent Raman conversion and walk-off. For pulse compression the red shifted spectral region which also corresponds to the part of the pulse nearly linearly chirped was selected using an asymmetric spectral window. In this manner wing-free compressed pulses were obtained which are stabilized against pump pulse fluctuations.

Now let us turn to the features of the Stokes pulse. Measurements of the pulse shape of Stokes pulses, their spectrum and chirp for ps pump pulses were reported in [4.117], [4.118]. From the measured delay between pump and Stokes pulse, the duration of the Stokes at the fiber output, and the known dispersion parameters of the fiber one can conclude about the origin of the Stokes pulses (which differs from the fiber entrance as already mentioned) and the initial Stokes duration. It could be shown that the distance between entrance and apparent pulse origin decreases with increasing Raman gain [4.117]. The pulse duration measured after various fiber lengths for equal input parameters indicated a strong chirp of the

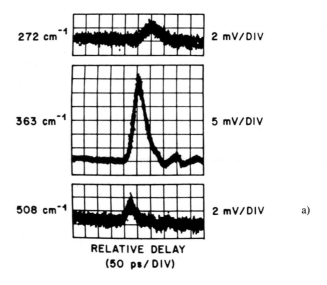

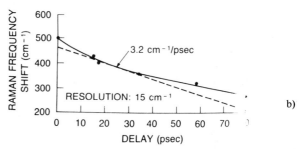

**Figure 4.23** Direct measurement of Raman pulse chirp, a. Sampling oscilloscope response of the dispersed Raman spectrum for different frequency shifts from the pump, b. Measured chirp of the Stokes pulse (bandwidth-limited pump pulses at $\lambda_L = 532$ nm and $L_F = 21$ m, $d = 4.1$ $\mu$m) (from [4.118]).

Stokes pulse at the location of its origin. For a direct chirp measurement the spectrum of the Stokes pulse was dispersed by a grating and its parts were imaged onto a high-speed photodetector. Figure 4.23a shows different parts of the Raman spectrum as function of the relative delay to the pump pulse from which the temporal behavior of the Raman frequency shift can be determined, see Fig. 4.23b. It is interesting to note that the measured chirp $d\nu/dt$ ($\nu = \omega/2\pi$) is almost constant and amounts to $9.6 \cdot 10^{10}$ Hz/ps which is larger by a factor 3 than the chirp of the pump at the exit of the fiber. These up chirped Stokes pulses can be compressed by ordinary grating compressors for example.

### 4.3.2   Influence of SRS on soliton propagation

As already discussed in section 4.1.5.2 SPM and anomalous GVD can result in the formation of solitons which are characterized by distortionless pulse propagation ($N = 1$ solitons) or which exhibit a periodic behavior ($N \geqslant 2$ solitons) while travelling through the fiber.

Considering the $N = 1$ soliton, SRS does not disturb the existence of the soliton as a stable entity [4.110]. However, in [4.119] it was discovered that the soliton propagation is accompanied by a continuous shift to lower frequencies — the so-called soliton self-frequency shift — which is the result of a continuous transfer of energy from higher to lower frequencies within the soliton spectrum due to SRS. Interesting effects occur if pulses are launched into the fiber the power of which exceeds the value corresponding to the $N = 1$ soliton. The generation of high energy solitons through SRS was proposed in [4.125] and first experimentally verified in [4.114], and can be understood as follows. The pulse narrowing taking place immediately after the fiber input is now accompanied by SRS leading to the formation of Stokes pulses which in turn also undergo soliton shaping. Utilizing these effects various groups reported on the generation of soliton pulses of 55-400 fs duration in the wavelength region 1.36–1.54 $\mu$m [4.124] starting from a synchronously pumped dye laser (0.83 ps, 1.341 $\mu$m, 70 mW, 82 MHz), on 80-200 fs Raman pulses obtained from 100 ps Nd:YAG laser pulses at 1.32 $\mu$m [4.122] and on the generation of 70 fs pulses at 1.6 $\mu$m using the 150 ps pulses from a Nd:YAG laser (1.06 $\mu$m) as the pump [4.123]. Note that the latter is an example of a cascade process, namely the shaping of higher order Stokes components. It is also noteworthy that in [4.124] the authors failed to observe solitons of higher order, instead, the input pulse breaks up temporally and spectrally after an initial stage of narrowing and one or more ultrashort Stokes pulses are formed corresponding to fundamental solitons. The frequency of these pulses continuously shifts to lower frequencies during the propagation through the fiber. Figure 4.24 shows the duration and the self-frequency shift of the Raman pulse as a function of the propagation distance through the fiber. It should be mentioned that most of the experiments described were performed with pump pulses the wavelength of which was in the neighbourhood of the zero dispersion point. That means the pump pulses can not undergo soliton shaping whereas the longer-wavelength Stokes pulses do.

### 4.3.3   Description of the propagation of Stokes and pump pulses

The basic equations considered so far (see (4.4)) have to be extended in order to describe the interaction between pump and Raman pulses in

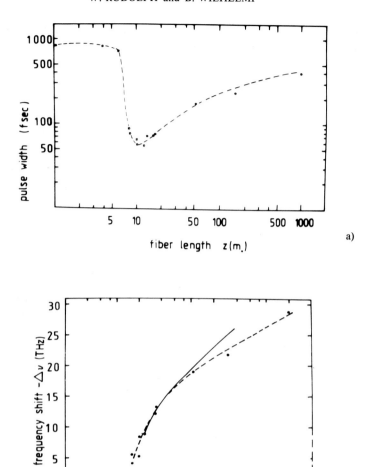

**Figure 4.24** Stokes pulse parameters as a function of the fiber length, a. duration, b. frequency shift, The solid line corresponds to theory. ($\lambda_L = 1.341$ $\mu$m, $\tau_L = 0.83$ ps, $P_L = 530$W) (from [4.124]).

nonlinear dispersive media. We take into account Raman-type and Kerr-type nonlinearities of third order, i.e.

$$P^{NL} = P_{Raman} + P_{Kerr} \tag{4.57a}$$

where $P_{Kerr}$ is assumed to be dispersionless in the frequency range considered. Now equation (4.3a) can be rewritten as

$$\bar{P}_{Kerr}(\omega_S,t) = 2\epsilon_0 n\bar{n}_2(|\bar{E}_L|^2 + |\bar{E}_S|^2)\,\bar{E}_S \qquad (4.57b)$$

and

$$\bar{P}_{Kerr}(\omega_L,t) = 2\epsilon_0 n\bar{n}_2(|\bar{E}_L|^2 + |\bar{E}_S|^2)\,\bar{E}_L \qquad (4.57c)$$

where the index S(L) stands for Stokes (pump) pulse. Regarding the Raman process, the material is described by interaction-free two-level systems with resonance frequency $\omega_{21}$ and phase decay time $\tau_{21}$. This is of course a rough approximation; the essential features can, however, be explained in this way. With neglect of the occupation of the higher energy level, which is justified for typical conditions, the equation of motion for the off-diagonal matrix amplitude $\bar{S}_{12}$ reads (compare, e.g., [4.137])

$$\frac{\partial}{\partial t}\,\bar{S}_{12} + \frac{1}{\tau_{21}}\,\bar{S}_{12} - i(\omega_{21} - \omega_L + \omega_S) = \frac{i}{2\hbar}\,\alpha_{21}\,\bar{E}_L^*\,\bar{E}_S e^{i\Delta\vec{k}\vec{r}} \qquad (4.58a)$$

where $\Delta\vec{k} = -\vec{k}_L + \vec{k}_S$; $\alpha_{21}$ is the Raman polarizability. In the following we set $\omega_S = \omega_L - \omega_{21}$, that means any deviations of the mean frequency of the Stokes pulse from $\omega_L - \omega_{21}$ appear as a phase modulation in $E_S$.

The evolution of the shape of the Stokes pulse under the action of both types of nonlinearities is now described by the following wave equations:

$$\left(\frac{\partial}{\partial z} + \frac{1}{v_S}\frac{\partial}{\partial t} - \frac{i}{2}\,k_S''\frac{\partial^2}{\partial t^2}\right)\bar{E}_S = \frac{\mu_0 N\alpha_{21}}{4ik_S}\,\bar{S}_{21}\,\bar{E}_L\,e^{-i\Delta\vec{k}\vec{r}}$$
$$- i\tilde{\kappa}_2^S(|\bar{E}_L|^2 + |\bar{E}_S|^2)\bar{E}_S \qquad (4.58b)$$

$$\left(\frac{\partial}{\partial z} + \frac{1}{v_L}\frac{\partial}{\partial t} - \frac{i}{2}\,k_L''\frac{\partial^2}{\partial t^2}\right)\bar{E}_L = \frac{\mu_0 N\alpha_{21}}{4ik_L}\,\bar{S}_{12}\,\bar{E}_s\,e^{i\Delta\vec{k}\vec{r}}$$
$$- i\tilde{\kappa}_2^L(|\bar{E}_L|^2 + |\bar{E}_S|^2)\bar{E}_L \qquad (4.58c)$$

where N is the number density of effective two-level systems and the coupling parameters $\tilde{\kappa}_2^S$, $\tilde{\kappa}_2^L$ are proportional to $\bar{n}_2$ (cf. equation (4.4b)). (Note that the real part of the third order nonlinear susceptibility describes SPM, the imaginary part is responsible for Raman scattering). Equation (4.58a–c) must now be solved simultaneously. As this task is rather tedious, we restrict ourselves in the following to several limiting cases.

*4.3.3.1 Formation of chirp in the Stokes pulses.* We will deal with the formation of chirp with rather long input pump pulses. First we consider the stationary solution $\left(\dfrac{\partial}{\partial t}\,\bar{S}_{12} = 0\right)$ of (4.58a) and use it in (4.58b,c). In

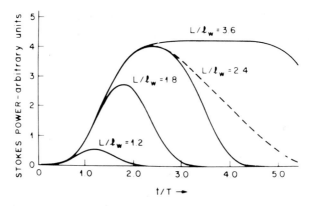

**Figure 4.25** Calculated instantaneous Raman Stokes power versus time for different fiber lengths neglecting power depletion (dashed line — with depletion) (from [4.117]).

this case the Raman process itself does not contribute to the chirp production but its peculiarities may strongly influence the shape of the Stokes and pump pulses. The pulse shapes, on the other hand, affect the chirp production via the Kerr terms in (4.58b,c). In discussing the original formation of the Stokes pulse, the dispersion broadening of laser and Stokes pulses can often be neglected as the main process takes part on some walk-off lengths $l_w$, which are small compared to $L_D$.

From (4.58b,c) we derive equations for the pulse envelopes or intensities and the time dependent phases, where the intensity equations are independent of the phases and will be treated first. For small Stokes conversion and normal dispersion the Stokes pulse duration increases monotonously with increasing fiber length, and the pulse becomes almost rectangular because the Stokes light generated first outruns the pump pulse while new noise at the Stokes frequency is amplified by its interaction with the (undepleted) laser pulse. This situation changes when the depletion of the laser pulse plays an essential part. The results are qualitatively demonstrated in Fig. 4.25. In agreement with experiment the Stokes pulse may be shorter than the input pulse at high conversion rates. The phase modulation of Stokes and pump pulses can now be calculated using the results for the intensities $I_L$ and $I_S$. The chirp of the Stokes pulse originates from the Kerr effect induced by both the pump pulse and the Stokes pulse. The first one acts only over a short piece of fiber, where both pulses overlap significantly.

Next we consider the influence of noninstantaneous response in a rough approximation. We neglect the pump depletion, which is justified for $|\bar{E}_S|$

$\ll |\overline{E}_L|$. In this case (4.58c) is independent of the other equations; it represents the NLSE for the pump pulse and can be solved as described in section 4.1. Now (4.58a) is solved for $\overline{S}_{21}$, which is substituted into (4.58b). The result is

$$\frac{\partial^2 \overline{E}_S}{\partial \zeta \partial \eta} + D_w \frac{\partial^2 \overline{E}_S}{\partial \eta^2} + \left(\frac{1}{\tau_{21}} - \psi\right)\left[\frac{\partial \overline{E}_S}{\partial \zeta} + D_w \frac{\partial \overline{E}}{\partial \eta}\right]$$

$$- \frac{i}{2} k_S'' \left[\frac{\partial^3 \overline{E}_S}{\partial \eta^3} + \left(\frac{1}{\tau_{21}} - \psi\right)\frac{\partial^2 \overline{E}_S}{\partial \eta^2}\right]$$

$$+ i\tilde{\kappa}_2^S |\overline{E}_L|^2 \left[\left(\psi^* + \frac{1}{\tau_{21}}\right)\overline{E}_S + \frac{\partial}{\partial \eta} \overline{E}_S\right] - \frac{Q}{\tau_{21}} |\overline{E}_L|^2 \overline{E}_S = 0 \quad (4.59)$$

where we used the local coordinates referring to the Stokes pulse $\eta = t - z/v_S$, $\zeta = z$ and the abbreviations $D_w = \frac{1}{v_S} - \frac{1}{v_L}$, $\psi = \frac{\partial}{\partial \eta} \ln$ $\overline{E}_L(\eta)$ and $Q = \alpha_{21} \mu_0 N \omega_S^2 \tau_{21}/(8\hbar k_S)$. $\overline{E}(\eta,\zeta) = |\overline{E}(\eta,\zeta)| \exp(i\Phi(\eta,\zeta))$ is a solution of (4.58c). This equation has been solved in [4.129] for several cases. Here in particular the limiting cases of quasi-stationarity ($\tau_{21} \ll \Delta\omega_L^{-1}$, $\Delta\omega_S^{-1}$) will be dealt with. Furthermore, we assume the length $L_F$ of the fiber to be short compared to the pulse dispersion length $L_D$. Using the ansatz $\overline{E}_S = A_S(\eta,\zeta)\exp(i\Phi(\eta - D_w\zeta, \zeta))$ and neglecting the terms proportional to $k_S''$ one obtains

$$\left[\frac{\partial}{\partial \eta} + \frac{1}{\tau_{21}} + i \frac{\partial}{\partial \eta} \Phi_L(\eta - D_w\zeta, \zeta) - \frac{\partial}{\partial \eta} \ln \overline{E}_L\right]\left(\frac{\partial}{\partial \zeta} A_S + D_w \frac{\partial}{\partial \eta} A_S\right)$$

$$= |\overline{E}_L(\eta)|^2 \left\{\frac{Q}{\tau_{21}} - i\tilde{\kappa}_2^S \left[\frac{\partial}{\partial \eta} + i \frac{\partial}{\partial \eta} \Phi_L(\eta - D_w\zeta, \zeta)\right.\right.$$

$$\left.\left. + \frac{\partial}{\partial \eta} \ln \overline{E}_L^* + \frac{1}{\tau_{21}}\right]\right\} A_S \quad (4.60)$$

By self-phase modulation the phase of the laser pulse evolves near the pulse peak as

$$\Phi_L(\eta,\zeta) = (\beta_0 + \beta_L\zeta)\eta^2 \quad (4.61)$$

where $\beta_0 = 0$ for bandwidth-limited entrance pulses. As we are interested in the formation of strongly chirped light pulses we assume $|\beta_L\xi| \gg 1/\tau_L^2$ for the main part of the interaction length. Then the time derivative of $\ln$ $\overline{E}_L(\eta)$ can be approximated by that of $i\Phi_L(\eta)$ and we have

$$i \frac{\partial}{\partial \eta} \Phi_L(\eta - D_w\zeta, \zeta) - \frac{\partial}{\partial \eta} \ln \overline{E}_L(\eta,\zeta) = -i2\beta_L D_w\zeta \quad (4.62)$$

84        W. RUDOLPH and B. WILHELMI

Using this equation and the approximation of quasi-stationarity we arrive at

$$\left(\frac{\partial}{\partial \zeta} + D_w \frac{\partial}{\partial \eta}\right) A_S = \left(\frac{Q}{1 - 2\beta_L D_w \zeta^2 \tau_L} - i\tilde{\kappa}_2^S\right) |\bar{E}_L|^2 A_S \quad (4.63a)$$

the solution of which is

$$A_S(\eta,\zeta) = A_{S0} \exp\left\{\int_0^{L_F} d\zeta' |\bar{E}_L(\eta + D_w\zeta', \zeta')|^2 \times \right.$$

$$\left. \times \left(\frac{Q}{1 - 2\beta_L D_w \zeta'^2 \tau_L} - i\tilde{\kappa}_2^S\right)\right\} \quad (4.63b)$$

The evaluation of (4.63b) for a Gaussian-shaped pump pulse

$$|\bar{E}_L(\eta)|^2 = A_0^2 \exp\left(-\frac{4\ln 2}{\tau_L^2}\right) \quad (4.64)$$

leads to

$$A_S = A_{S0} \exp\left\{\tfrac{1}{2}G - \frac{2\ln 2}{\tau_L^2}(\eta - \eta_0)^2 + i\Delta\omega_S(\eta - \eta_0) \right.$$

$$\left. + i\beta_S(\eta - \eta_0)^2\right\} \quad (4.65a)$$

where

$$G = \frac{\sqrt{\pi}\,\tau_L A_0^2 Q}{\sqrt{\ln 2}\,(-D_w)}\,\mathscr{F}\left(-\frac{D_w L_F}{\tau_L}\sqrt{2\ln 2}\right) \quad (4.65b)$$

is the Stokes gain with $\mathscr{F}$ being the error function.

$$\tau_S = \frac{\tau_L}{\sqrt{2QL_FA_0^2}}\exp\left\{\frac{L_F^2 D_w^2 \ln 2}{2\tau_L^2}\right\} \quad (4.65c)$$

is the Stokes pulse duration, and

$$\beta_S = \beta_L L_F + \frac{4\ln 2\,\tilde{\kappa}_2^S A_0^2 L_F}{\tau_L^2}\exp\left\{-\frac{D_w^2 L_F^2}{\tau_L^2}\ln 2\right\}$$

$$+ QA_0^2\beta_L D_w \tau_{21}\left\{\frac{\sqrt{\pi}\,\tau_L}{D_w^3\sqrt{\ln 2}}\,\mathscr{F}\left(-\frac{D_w L_F}{\tau_L}\sqrt{2\ln 2}\right)\right.$$

$$\left. + \frac{L_F}{D_w^2}\left(2 + \frac{L_F^2 D_w^2 4\ln 2}{\tau_L^2}\right)\exp\left[-\left(\frac{D_w L_F}{\tau_L}\sqrt{\ln 2}\right)^2\right]\right\} \quad (4.65d)$$

is the chirp parameter of the Stokes pulse (note that $D_w < 0$ for normal dispersion). The first term in (4.65d) represents the direct transformation of chirp from the laser wave to the Stokes wave. The second term is due to the Kerr effect, which the Stokes wave experiences in the field of the laser wave; it approaches zero for $|D_w|L_F \gg 1$ because the Stokes pulse increases significantly in duration as it is seen from (4.65c). The third term gives the first nonstationary correction, it is obviously caused by the interplay between nonlinearity ($\beta_S \propto A_0$), transient response ($\beta_S \propto \tau_{21}$) and dispersion ($\beta_S \propto D_w$). For $|D_w|L_F/\tau_L > 1$ the gain is limited by pulse walk-off and $\tau_L/|D_w|$ acts as the effective interaction length. The maximum gain is given by

$$G_m = \frac{\sqrt{\pi}}{\sqrt{\ln 2}}\,\frac{\tau_L}{|D_w|}A_0^2 Q \simeq 1.06 g_0\,I_0\frac{\tau_L}{|D_w|} \qquad (4.66)$$

where $I_0$ is the initial laser power per area and $g_0$ is the gain coefficient for stationary, dispersionless Stokes pulse amplification. Then the ratio between the additional chirp of the Stokes pulse and that of the laser pulse is

$$\frac{\Delta\beta_S}{\beta_L} \simeq \left(1.06 g_0\,I_0\,\frac{\tau_L}{|D_w|}\right)\left(\frac{\tau_{21}}{\tau_L}\right)\left(\frac{\tau_L}{|D_w|L_F}\right) \qquad (4.67)$$

An order of magnitude estimation shows the $\Delta\beta_S/\beta_L$ may attain values in the order of one. Thus, this term might also be responsible for some part

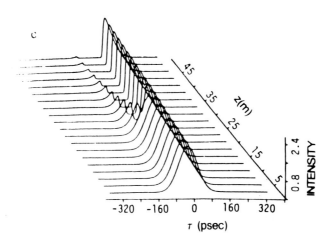

**Figure 4.26** Calculated temporal shapes of pump pulses as a function of propagation distance ($\tau_L = 100$ ps, $P_L = 1.75$ kW). The Stokes pulse is not shown here (from [4.130]).

of excess chirp of the Stokes pulse measured in [4.118]. For quantitative comparison one has to solve (4.58) numerically without the use of the small signal approximation (the Stokes conversion attains values of about 30%) and without the assumption of quasi-stationarity (the spectral width of the Stokes pulse approaches the order of $1/\tau_{21}$). Such nonstationary corrections become obviously more important for shorter input pulses. A numerical evaluation of (4.58) for the stationary case $\left(\dfrac{\partial}{\partial t}\,\overline{S}_{21}=0\right)$ and assuming $k''_{L,S}=0$ was carried out in [4.130]. Results concerning the temporal shapes of the pump pulse as a function of fiber length are shown in Fig. 4.26. Pump pulse depletion, pulse fragmentation and the influence of walk-off can clearly be seen.

### 4.3.3.2  Propagation of very short light pulses in fibers. Mean frequency shift of solitons.

Next we treat the effect of the Raman-type nonlinearity on the propagation of very short light pulses in fibers with $k''_L\,\tilde{\kappa}_2<0$ in particular on the propagation of solitons. Such pulse propagation was treated by solving the NLSE. This means the third order nonlinearity is assumed to act instantaneously. In general, however, the third order polarization is given by

$$\overline{P}^{NL}(t) = \epsilon_0 \int\limits_0^\infty d\tau_1 \int\limits_0^\infty d\tau_2 \int\limits_0^\infty d\tau_3 \, \hat{X}^{(3)}(\tau_1,\,\tau_2,\,\tau_3)\, E(t-\tau_1)\, E(t-\tau_2)\, E(t-\tau_3)$$

$$(4.68a)$$

(see, e.g., [4.137]).

When the Raman process is mainly responsible for the delayed response (this holds when the pulse frequencies are far from one-photon resonances) the simpler relation

$$\overline{P}^{NL}(t) = \overline{E}(t) \int\limits_{-\infty}^\infty dt'\hat{X}^{(3)}(t')|\overline{E}(t-t')|^2 \qquad (4.68b)$$

can be used. This term can be obtained by integration of (4.68a) with $E_S = E_L = E$. When the material response is fast ($\tau_{21} \ll \Delta\omega^{-1}$), equation (4.68b) can be expanded in the form

$$\overline{P}^{NL}(t) = X^{(3)}E(t) \left[ |\overline{E}(t)|^2 - c_1\,\frac{\partial}{\partial t}\,|\overline{E}(t)|^2 + \frac{1}{2}\,c_2\,\frac{\partial^2}{\partial t^2}\,|\overline{E}(t)|^2 + \ldots \right]$$

$$(4.68c)$$

where the first term describes the instantaneous response and the next terms — deviations from it (see, e.g., in [4.137] p. 74). In [4.127] the influence of such additional terms on the propagation of solitons in fibers was discussed. The authors considered the additional terms as a small perturbation for the solitons, which are obtained by solving the ordinary NLSE. The Raman-type nonlinearity causes losses at the high-frequency side of the pulse spectrum and gain at the low-frequency side. This causes a shift of the mean frequency $\omega_0$ of the light pulse with increasing fiber length z which is calculated to be

$$\frac{d\omega_0}{dz} = -\frac{\pi}{8}\frac{\lambda_0}{2\pi\bar{n}_2}\int d\tilde{\Omega}\ \tilde{\Omega}^3 g(\tilde{\Omega})\ \sinh^{-2}(\pi\tilde{\Omega}/2) \qquad (4.69a)$$

where $\tilde{\Omega} = 0.56\ \omega\tau_L$ ($\tau_L$: FWHM duration of the soliton, $g(\tilde{\Omega})$ is the Raman gain coefficient). An evaluation for quartz fibers at a wavelength of 1.5 $\mu$m yields

$$\frac{d\omega_0}{dz}\ [\text{THz/km}] \simeq 0.27/[\tau_L/\text{ps}]^4 \qquad (4.69b)$$

where as a rough approximation $g(\tilde{\Omega}) \propto \tilde{\Omega}$ has been used. A more detailed discussion of the integral is to be found in [4.127], but even this simple relation explains at least qualitatively the experimental results described before (see Fig. 4.24). In order to study soliton shaping and SRS more quantitatively various numerical treatments of the extended NLSE were performed [4.125], [4.128], [4.131] by which means most of the experimental findings described here could be explained.

## 5.  INTRACAVITY PULSE COMPRESSION

While in extracavity pulse compression the processes of chirp generation and compensation can clearly be distinguished the situation is more complex for intracavity pulse compression. The development of lasers emitting pulses with durations on a femtosecond time scale made it necessary to combine phase modulation processes resulting from the various nonlinear optical processes as well as linear dispersive effects from the resonator elements with the pulse shaping in the saturable media for describing the experimental findings. It turned out that the influence of linear dispersion in the cavity (from mirrors, glass paths etc.) which can be neglected in good approximation in most picosecond and nanosecond lasers, has to be considered in dealing with femtosecond lasers where the pulse bandwidths are on the order of several nanometers.

On the one hand, care has to be taken to avoid net GVD in the resonator yielding to an undesired pulse spreading. On the other hand linear dispersion can be tuned to compensate for phase modulation arising from nonlinear processes which may lead to an additional pulse shortening. Of course, this chirp generation and compensation proceeds simultaneously with other linear and nonlinear processes and cannot be discussed independently from them.

The generation of a nonlinear phase modulation in the laser cavity was observed in solid state lasers many years ago and effort was made to compress the pulses outside the cavity [5.1]. Also, the pulses from a passively mode-locked and cavity dumped dye laser turned out to be phase modulated [5.2] which gave rise to a pulse shortening after suitable spectral filtering.

With the introduction of the Colliding-Pulse-Modelocking (CPM) [5.3] in dye lasers, it was possible for the first time to produce stable trains of pulses shorter than 100 fs. So it became possible to investigate and to utilize intracavity self-phase modulation and pulse compression [5.4–5.9]. Some of the experimental results will be discussed in Section 5.1.

The theoretical description of these lasers utilizes the fact that the steady state pulse regime, in which the chirp generation and compensation as well as the shaping due to saturable absorption and amplification counterbalance each other during one cavity round trip or after a certain number of round trips, can be thought of as soliton-like pulse propagation [5.10–5.13]. We will describe its main principles in Section 5.2.

While this type of intracavity pulse shaping rests on an additional adjustable element for GVD (in analogy to the extracavity fiber grating compressor) there are also lasers where the soliton shaping mechanism occurring in suitable nonlinear, dispersive samples (fibers) is directly used intracavity for producing fs pulses. This so-called "soliton laser" [5.14], [5.15] will be dealt with in section 5.3.

## 5.1 Experimental

Because of their large amplifying bandwidth, dye lasers have been most advantageously used for producing pulses with durations well below 100 fs [5.16–5.25], [5.3], [5.4]. As a consequence the effects of intracavity GVD and SPM become particularly important in such lasers and have been extensively studied.

In the last few years several cavity configurations and modelocking

schemes have been developed for generating fs light pulses. Among them
are ring resonators [5.3], [5.4], [5.6], [5.9], [5.17], [5.20], [5.21], [5.26],
linear resonators [5.27], [5.28], [5.22], [5.24] and linear resonators terminated with an antiresonant ring [5.18], [5.19], [5.23] which are passively
[5.3], [5.4], [5.6], [5.9], [5.17], [5.21], [5.24], [5.29] or hybridly [5.16],
[5.20], [5.22], [5.26] modelocked. By using different combinations of
absorber and amplifier dyes it was also possible to extend the wavelength
range for stable femtosecond pulse generation up to the near infrared
[5.20], [5.24].

Most of these lasers have special optical components for aligning the
intracavity GVD with the aim to achieve the shortest possible pulses in
common. Here we want to discuss some basic experimental results
obtained with the passively mode-locked CPM ring laser, which so far is
the most successful and widely used concept for the design of fs lasers.

The CPM ring laser is essentially a passively mode-locked laser consist-
ing of a depletable amplifier and a saturable absorber, where the collision
of two counterpropagating pulses in the absorber modifies the saturation
behavior of the absorber and stabilizes the pulse generation. Mostly Rh6G
and DODCI are used as amplifier and absorber which determines the laser
wavelength to be between 605 nm and 635 nm. Figures 5.1, 5.2 show
typical ring resonator configurations by means of which it was possible to
produce pulses of about 30 fs. Critical points of the resonator alignment
are suitable focusing in the absorber and amplifier jet to provide a certain
ratio of the intensities in the saturable media, which is particularly
difficult because the passive resonator exhibits several stability regions

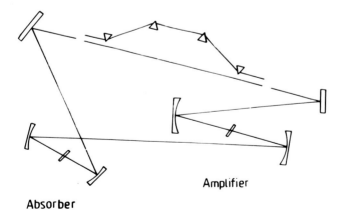

Amplifier

Absorber

**Figure 5.1**    Ring resonator configuration for a CPM dye laser. With the insertion of a four-
prism sequence the intracavity GVD can be controlled (from [5.3], [5.6]).

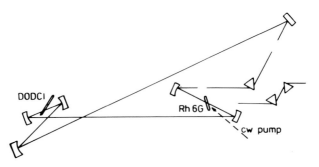

**Figure 5.2** Ring resonator configuration for CPM dye lasers. One prism or a two-prism sequence are for the GVD adjustment (from [5.29], [5.4], [5.5]).

separated from each other (e.g. there are several combinations of the distances between the two focusing mirror pairs possible). Moreover, in order to avoid spectral narrowing as well as undesired GVD, mirrors with broad constant reflecting band should be used.

Originally, the CPM laser [5.3] was designed without the possibility for adjusting the intracavity GVD (resonator configuration according to Fig. 5.1 without prism) and the combined influence of mirrors (see Section 3), nonlinear media, etc. produced a net phase modulation. So the pulses at the output of the laser (Fig. 5.2, one prism) were proved to be downchirped [5.30] by demonstrating their compressibility when traversing through a material with positive GVD (glass) outside the resonator, see Fig. 5.3.

The role that intracavity GVD plays in the lasers was demonstrated in [5.4]. Here the pulse duration was measured as a function of the amount of intracavity GVD (Fig. 5.4). The GVD (intracavity glass path) could be continuously changed by translating the intracavity prism normally to its base. At an optimum glass path, i.e., maximum possible intracavity pulse compression, pulses as short as 55 fs were observed that were almost bandwidth limited. The latter indicates an optimum compensation of the chirp generated during one cavity round trip. The ratio of extracavity and intracavity dispersion needed for chirp compensation and maximum pulse compression corresponds roughly to the number of cavity round trips within the resonator life time [5.10].

Improvements in the laser design and optimization of all cavity elements led to a further reduction of the pulse duration below 30 fs [5.6], [5.31a] and recently, the generation of 19 fs pulses [5.31b] was reported. The lasing wavelength of these lasers was shifted to the red (630 nm–635 nm)

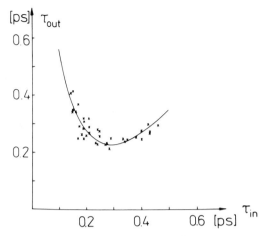

**Figure 5.3** First demonstration of the phase modulation of pulses from the CPM laser (according to Fig. 5.2 one prism). The pulses were sent through a 17 cm sample (BK7 glass). The different input pulse durations ($\tau_{in}$) were achieved by a small adjustment of the absorber jet while keeping the adjustment of the prism and mirrors unchanged. Long input pulses are compressed during their passage through the glass indicating down-chirp at the entrance of the sample. The broadening of shorter pulses is because the sample length is larger than $L_\beta$ (see 3.3c) (from [5.30]).

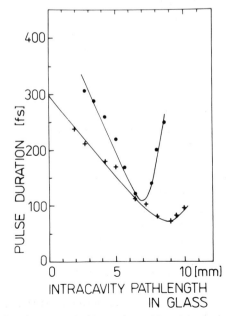

**Figure 5.4** Pulse duration versus the intracavity pathlength in glass for two different laser wavelengths and glass materials ( + : 615 nm, SQ1,.: 619 nm, Flint) (from [5.4]).

as opposed to the earlier configurations and lies on the long-wavelengths side of both and DODCI ground form and of its photoisomer.

For the proper adjustment of the intracavity dispersion, use was made of a four-prism sequence, a two-prism sequence and a single prism, respectively (Fig. 5.1, 5.2). It should be mentioned that there are also successful attempts to use a Gires-Tournois interferometer for this purpose [5.9]. Moreover, a curve similar to Fig. 5.4 could be measured by varying the intracavity dispersion through the exchange of mirrors exhibiting different $\psi(\omega)$ (see (2.22)) [5.7]. The dependence of the pulse duration on the intracavity GVD shows an asymmetric behavior, where the stable pulse regime often breaks up suddenly if positive GVD is added at the point of optimum compression. Note that this behavior is completely different from the extracavity compression experiments and indicates the strong interplay of the various linear and nonlinear processes contributing to the pulse evolution in the laser. In [5.8] it could be shown that even small changes of $\psi(\omega)$ by adjustment of suitable intracavity elements influence the pulse stability.

So far there are no direct measurements that indicate what kind and which amount of phase modulation is induced by the various resonator components. Also the net GVD at the point of maximum intracavity compression cannot be determined with certainty. As pointed out in Section 3 the amount of GVD arising from various linear elements can be estimated only under assumptions which are not completely valid in the resonator. The contribution of mirror dispersion can probably be neglected if the mirrors are suitably chosen [5.32], [5.6] (single stack multilayer mirrors with a center frequency near the laser wavelength and with a broad region of almost constant reflectivity). There remained many open questions, they concern, e.g., the effect of filters and the interplay between angular dispersion and image-forming elements (e.g. curved mirrors). (First attempts to describe the latter theoretically are found in [5.33], [5.34].)

That nonlinear processes play a part in the chirp formation can be seen from Fig. 5.5; similar investigations were also performed in [5.6]. Here the change of the intracavity glass path was measured as a function of the position of the absorber relative to the beam waist, which is a rough measure of the intensity in the absorber [5.5] and accordingly of the efficiency of the nonlinear processes (saturation, intensity dependent refractive index changes of the solvent) taking place there.

It should be noted that it is mostly impossible, or at least extremely difficult, to measure the functional dependence between two laser parameters while keeping all other parameters constant or even to repeat the measurements under equal conditions because of the large number of

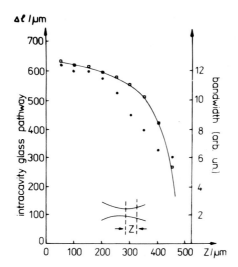

**Figure 5.5** Change of the intracavity glasspath (□) and the pulse bandwith (·) as a function of the deviation of the absorber position from the beam waist (from [5.5]).

degrees of freedom (e.g. resonator alignment, charge and age of the dyes used). Detailed experimental studies describing basic features of the CPM laser are found in [5.5], [5.6], [5.10], [5.3].

## 5.2 Theoretical

The intracavity chirp generation, chirp compensation, pulse compression and the other pulse shaping mechanisms proceed simultaneously during the pulse formation and are related in a complicated manner. Consequently, the intracavity pulse compression through chirp generation and compensation cannot be treated separately from the other processes, i.e., the complex theory of passive modelocking [5.35–5.44] had to take into account these processes. From [5.35–5.44] the basic principles of passive modelocking can be summarized roughly in the following way:

When the pump radiation exceeds the threshold for a large number of modes, the laser intensity is a statistical distribution of fluctuation peaks. Since the average duration of fluctuation spikes and that of the pulses formed later on in the evolution process are small compared with the fluorescence life times $T_1^\alpha$, $T_1^g$ of the absorber and gain medium, the saturation is controlled by the pulse energy. The absorber favors fluctuation peaks or groups of them with high energy as opposed to the depletable amplifier. For active medium and absorber with given interaction cross sections the relation between the two nonlinear processes can be adjusted by the beam

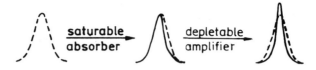

**Figure 5.6** Schematic of pulse shaping by passing through a saturable absorber and a depletable amplifier.

cross sections, to which the pulses are focused into the media. When the nonlinearity of the absorber is adjusted to be higher than that of the amplifier the peaks with the highest energy experience the highest net gain in each cavity round trip while the remaining ones are gradually suppressed. In this regime the absorber attenuates the leading edge of the pulse until the energy that has already passed the medium attains a value which is sufficient to modify (lessen) the absorption considerably. In order to generate stable ultrashot light pulses it is necessary to suppress not only the leading edge, but also the trailing edge of the pulse. This suppression is achieved due to gain depletion resulting from the decrease of the occupation number inversion during the transit of the pulse (gain saturation). With a certain choice of the laser parameters, in particular of the ratio between the round trip time U and the life time $T_1^g$ at given pump power, the pulse edges suffer net loss (Fig. 5.6). Within the stability range defined in this way the light pulses become shorter and shorter by the combined action of the amplifier and the absorber down to pulse durations where bandwidth-limiting effects counterbalance this pulse shortening. Under certain conditions a steady state is finally reached in which an ultrashort light pulse reproduces itself after each resonator round trip (steady state pulse regime). Note that according to our discussion in Section 4 the spectral profiles of absorber and amplifier remain as filters if all additional frequency selective optical components are carefully removed from the resonator.

With the aim to include pulse compression due to chirp generation and compensation in this concept, we have to modify the model discussed above with respect to two points. First, we have to account for the self-phase modulation associated with the interaction between the light pulse and the media (see Section 4). Second, an effective element representing linear dispersion (GVD) has to be introduced (see Section 3). Figure 5.7 shows the resulting resonator scheme to which we want to refer in the following.

The steady state pulse regime can advantageously be treated in the round trip model, where the requirement that the complex pulse amplitude

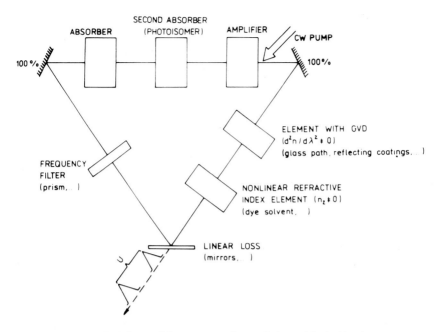

**Figure 5.7**   Scheme of the resonator of a passively modelocked laser.

has to reproduce itself after one passage through the cavity can be written
as

$$\overline{E}(\eta + h) = J_g\, J_a\, J_l\, J_{nl}\, \overline{E}(\eta) \tag{5.1}$$

$J_{g,a,l,nl}$ represent complex operators describing the action of the gain
medium (g), absorber medium (a), the linear element (l) and the nonlinear
refractive index element (nl); h is a possible shift of the pulse per round trip
in the local time $\eta$.

Note that in the general case, the pulse propagation through the media is
described by a system of partial differential equations (Section 4.2.1), and
the solution of the steady state relation (5.1) is rather difficult even if
numerical methods are applied. Therefore, one has to look for appropriate approximations which simplify the calculations but which, of
course, do not take into consideration the full complexity of the processes
taking place.

In [5.12], [5.46] equation (5.1) was evaluated within the conditions
given in [5.39], i.e., for small pulse energies as compared with the saturation energies of the transition and small gain and absorption. These
conditions are approximately met in weakly pumped systems. In addition,

the change of modulus and phase that the pulse suffers in the effective linear element (linear losses, GVD, frequency filter) are assumed to be small. Thus, we find approximate expressions for the operators J which are

$$J_{a,g} = 1 - \tfrac{1}{2}\kappa_{a,g}\mathscr{L}_{a,g}(1 - \bar{\epsilon}_{a,g}(\eta) + \tfrac{1}{2}\bar{\epsilon}_{a,g}^2(\eta)) \tag{5.2a}$$

$$J_l = 1 - \frac{2}{\Delta\omega_F}\frac{d}{d\eta} + \left(\frac{4}{\Delta\omega_F^2} + \frac{1}{2}ib_2\right)\frac{d^2}{d\eta^2} - \gamma \tag{5.2b}$$

$$J_{nl} = 1 - i\,\zeta_0\tilde{\kappa}_2\,|\bar{E}_0(\eta)|^2 \tag{5.2c}$$

Equation (5.2a) simply follows from (4.54) when the phase memory is neglected ($T_2 \rightarrow 0$).

$J_l$ is the product of a Lorentzian shaped filter function (see Table 3.1) which is to represent the bandwidth limiting properties of the cavity and the transfer function of a linear element exhibiting GVD (2.22d). Both functions have been expanded up to terms proportional to $(1/\Delta\omega_F\tau_L)^2$ and $(b_2/\tau_L^2)$ respectively, where $\Delta\omega_F$ is the bandwidth of the filter and $b_2 = (d^2\psi/d\omega^2)|_{\omega_L}$ stands for the dispersion, $\gamma$ accounts for linear losses. (5.2c) is obtained by expanding (4.10) under the condition of $\zeta_0\tilde{\kappa}_2|\bar{E}_0|^2 \ll 1$, where $\zeta_0$ is the length of the nonlinear medium. The use of (5.2a–c) in (5.1) yields an integro-differential equation with complex coefficients. In [5.45], [5.46] an approximate analytical ansatz for modulus and phase was found that solves this equation; it is

$$\bar{E}(\eta) = A(\eta)e^{i\Phi(\eta)} \tag{5.3a}$$

$$A(\eta) = A_0/\cosh(\eta/T) \tag{5.3b}$$

$$\frac{d}{d\eta}\Phi(\eta) = \frac{b}{T}\tanh(\eta/T) \tag{5.3c}$$

In this manner the mathematical problem can be reduced to a system of algebraic equations for the unknown quantities $\bar{\epsilon}_a(\infty)$ (the normalized pulse energy), $\omega_L$ (center frequency of the pulse spectrum), T (FWHM $\simeq$ 1.76T) and b (parameter describing the phase modulation). If we neglect the influence of the nonlinear refractive index element which is possible for relatively low intracavity laser intensity and relatively thin media ($\zeta_0 < 10^{-4}$ m) and assume that the net GVD of the cavity vanishes the chirp parameter is determined only by the absorber (if it is saturated much stronger than the gain medium) and is given by

$$b = -\frac{3}{2}\frac{L_r^a}{L_i^a} + c_0\left[\left(\frac{3L_r^a}{2L_i^a}\right)^2 + 2\right]^{1/2} \tag{5.4}$$

with $c_0 = \text{sgn}(\omega_a - \omega_L)$.

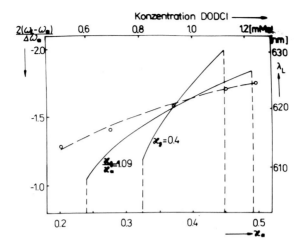

**Figure 5.8** Measured laser wavelength ( · ) as a function of the absorber concentration [5.4] and calculated normalized frequency as a function of the absorption coefficient $\kappa_a$ for constant amplification coefficient $\kappa_g$ and constant ratio of absorption and amplification coefficients $\kappa_a/\kappa_g$ (from [5.46]).

($\omega_a$: mid frequency of the absorbing transition), which results in a down chirp (b < 0) for $\omega_L < \omega_a$. The value of the dispersion parameter $b_2$ which gives the net GVD of the cavity necessary for complete chirp compensation is found to be [5.47]

$$b_2 \simeq \frac{5}{8} \kappa_a L_i^a \bar{\epsilon}_a^2(\infty) T^2 \qquad (5.5)$$

For typical parameters of a weakly pumped laser (T $\simeq$ 50 fs, $\bar{\epsilon}_a(\infty) = 0.6$, $\kappa_a L_i^a \simeq 0.2$) we can evaluate $b_2 \simeq 1.1 \ 10^{-28}$ s, which corresponds to about 2 mm intracavity glass (SQ 1) that would be necessary to compensate for the chirp generated in the absorber.

A closer inspection of the system of algebraic equations shows that the laser frequency $\omega_L$ can be tuned by changing the ratio of gain and absorption coefficient (see Fig. 5.8) which is in qualitative agreement with experiment and which in this parameter range can roughly be explained by the shift of the maximum of the small signal net gain.

At higher intracavity intensities additional processes have to be considered as the chirp generation in a nonlinear refractive index element [5.48], [5.12]. The sign of the chirp introduced by the solvent in this way is positive and opposite to that induced by the absorber.

Starting with the same analytical ansatz (5.3), these conditions were studied in [5.12], [5.49], where the resulting system of algebraic equations for the unknown quantities was solved under some restricting simplifications. It was found that due to the partial compensation of the two nonlinear self-phase modulation mechanisms the shortest achievable pulse duration taken at optimum chirp compensation increases with increasing influence of the Kerr nonlinearity. On the other hand, the pulses became shorter if the laser frequency was assumed to be at the short wavelength side of the absorbing transition, i.e., absorber and solvent produce chirp of equal sign (up chirp). This additional pulse shortening was on the order of some ten percent of the pulse duration.

A similar treatment was chosen in [5.50], where the effect of SPM due to a nonlinear refractive index element was calculated with respect to various types of modelocking. A more intuitive way was used in [5.51] to study the interplay of nonlinear optical self-phase modulation and GVD in a passively mode-locked laser. The steady state was assumed to be only determined by the requirement of a balance between pulse shortening and pulse broadening from which the unknown pulse duration and chirp parameter were derived.

With the aim to produce pulses as short as possible experimental effort has been made to avoid passive elements which act as filters and restrict the laser bandwidth. As already mentioned, in any case the saturable media remain as filters. To deal with this limiting case of passive modelocking the REA used so far is no longer applicable, but the bandwidth limiting properties of the media must be taken into consideration. Within the model used for the saturable media in section 4 this can in approximation be done by introducing a certain phase memory ($T_2 \neq 0$ in (4.54)) for the absorber and amplifier transition. The round trip equation resulting now from (5.1) and the modified equation (5.2) was solved in [5.11] to study the effect of phase modulation arising from the off-resonant interaction with the saturable media and subsequent complete chirp compensation (i.e. using an ideal compressor) on the pulse duration, see Fig. 5.9. For comparison, the pulse duration is also shown in Fig. 5.9 for a regime with resonant interaction (i.e. no production of chirp in the saturable media) which provides the same condition for the real amplitudes (i.e. absorption and gain coefficients, saturation, linear losses). The boundaries of the stability region were obtained from the requirement that the net gain at the leading and trailing edge of the pulse is negative. From this figure an additional shortening of about 1.2–2 can be expected.

As discussed above the finite width of the transitions of the amplifier and absorber set a limit for the achievable pulse duration. In the small signal case (i.e. without saturation) this is due to the spectral net gain profile acting as a linear filter. Shorter pulses should occur if the filter profile is broadened, for example, by inserting an additional filter with its

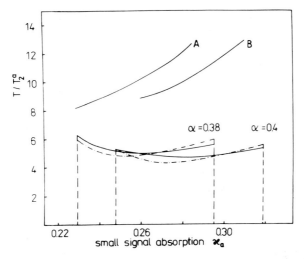

**Figure 5.9** Normalized pulse duration as a function of the small signal absorption for two values of the coefficient of the small signal amplification ($\alpha$) under complete chirp compensation. The wavelength was assumed to be at maximum net gain (—) and at maximum pulse energy (— · —), respectively. For comparison the pulse duration for the corresponding regimes (A,B) is shown, where no chirp is produced (from [5.11]).

transmission minimum at the frequency of the maximum of the net gain profile. Indeed it could be shown [5.11] that the pulses become shorter as the net gain profile flattens. At the point where the stable pulse solutions vanish, the profile is almost flat and the pulse duration is limited by the uncompensated dispersive properties of the saturated absorber and amplifier.

To deal passive modelocking without the restricting assumptions which were necessary for the more or less analytical treatments described above to be applied a numerical simulation was used in [5.52], [5.53] for solving (5.1). Instead of using (5.2a) for the light-matter interaction the system of differential equations (4.2), (4.45)–(4.47) was solved directly. The GVD element was taken into account through (2.22d) whereas (5.2c) remained unchanged. Starting from noise the temporal evolution of modulus and phase as well as the steady state were investigated under the presence of the resonator elements sketched in Fig. 5.7. After the pump is switched on, i.e. when the simulation has to start, the noise characterizes the spontaneous emission of the gain medium. Figure 5.10 shows an example of the temporal development of the pulse characteristics for a given set of laser parameters and with zero net GVD. The steady state is reached after several hundred round trips, where the pulse duration undergoes a monotonic decrease until its final value is reached.

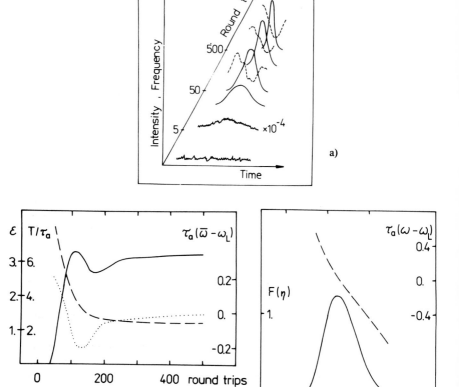

**Figure 5.10** Temporal evolution of the steady state in passively mode-locked dye lasers starting from noise, a. pulse modulus and frequency versus time for various values of the round trip number, b. normalized pulse energy $\epsilon$(——), pulse duration $T/\tau_a$ (---) and center frequency $\tau_a(\bar{\omega} - \omega_L)$ (. . .) versus round trip number (T: FWHM; $\tau_a = T_2^g$), c. steady state modulus (——) and instantaneous frequency (---) (from [5.53]).

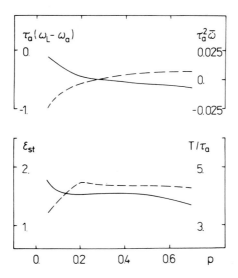

**Figure 5.11** Normalized mean laser frequency (——), mean chirp (---), pulse energy (——) and pulse duration (---) as a function of the photoisomer percentage p for a given set of laser parameters ($\tau_a = T_2^g$, normalized relative position of the transition maxima: $T_2^g(\omega_c - \omega_g) = -1$, $T_2^g(\omega_a - \omega_g) = -0.2$, $\omega_a \triangleq \omega_g$).

During the process of pulse evolution also the center frequency of the pulse approaches a steady-state value which mainly depends on the position of the transition maxima of absorber and amplifier. In this manner the occurrence of a photosisomer which is typical in using the saturable absorber DODCI [5.54], [5.55] affects the pulse center frequency in the steady state. The phase modulation now results from the interplay of three saturable media differing in the position of their transition profiles and saturation behaviors. Figure 5.11 shows the change of the pulse parameters when the percentage of the isomer (index c) is varied.

With the introduction of linear dispersion (GVD), i.e. $b_2 \neq 0$ in (2.22d), the pulse chirp can be shifted from negative to positive values where the pulses reach a minimum duration at a certain positive dispersion, Fig. 5.12. At this point of optimum intracavity compression, the pulses are slightly chirped.

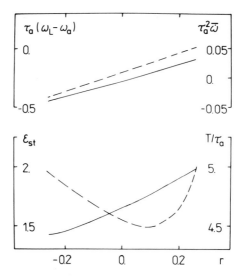

**Figure 5.12** Steady state pulse parameters (see Fig. 5.11) as a function of the intracavity GVD r. (r is proportional to $b_2$ from (2.22d)) (from [5.53]).

As outlined in section 4 in connection with extracavity compression a nonlinear refractive index element leads to the production of positive chirp associated with an increase of the pulse bandwidth, which can be utilized to compress the pulses through chirp compensation. In the resonator, however, the insertion of a nonlinear refractive index element does not necessarily result in an additional pulse shortening even if the GVD is suitably adjusted. For a given set of laser parameters there is always a combination of GVD and the value of the nonlinearity where the pulses become shortest, which is just at the boundary of the stability range, see Fig. 5.13. Larger values of the nonlinearity (here product of sample length and coefficient of the nonlinearity $\zeta_0 \tilde{\kappa}_2$, cf. (5.2c)) lead to instabilities (sometimes at least to periodically occurring satellites), where the reason is supposed to be that the GVD cannot compensate for the strong nonlinear chirp induced by the nonlinear refractive index element.

At this point it should be mentioned that as outlined in [5.10], [5.11], [5.13] the pulse propagation within the laser cavity can be thought of as soliton-like. In analogy to the soliton formation in optical fibers due to nonlinear self-phase modulation and simultaneously acting GVD, both processes also occur in the resonator, though, in combination with other pulse shaping mechanisms. In terms of the soliton propagation in fibers one (or possibly some) resonator lengths would correspond to the

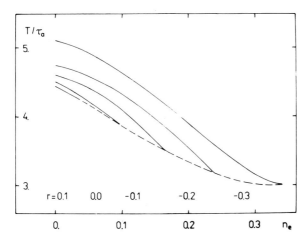

**Figure 5.13** Normalized pulse duration versus the normalized intensity-dependent refractive index $n_e$ for different values of the GVD r. ($\tau_a = T_2^g$, ---: boundary of the stable pulse regime, $n_e$ is proportional to $\bar{\kappa}_2 \zeta_0$) (from[5.56]).

periodicity length (see section 4.1.5.2) after which the pulse shapes are restored. There are interesting experimental hints supporting this concept [5.13]. However, so far there are no successful theoretical treatments of passive mode-locking making use directly of soliton solution methods (e.g. inverse scattering method) because of the large number of parameters resulting in a complicated system of differential equations. In the steady state pulse regime the laser can be considered as an infinitely extended medium composed of the resonator elements the properties of which are homogeneously distributed. Here the velocity of the form-stable pulse envelope would correspond to the soliton velocity. It should be mentioned that in the round trip model applied in [5.53] the steady state pulse parameters are independent of the special type of input signal; various types of noise as well as deterministic initial pulse have been used. To some extent this can also be regarded as an analogy to the formation of a soliton in fibers which develops from incident pulses, the parameters of which may differ to a great degree.

From the calculations interesting pulse solutions can be expected far above threshold. The pulse area (cf. (4.44)) approaches values where the interaction with the active media is coherent ($\pi$ pulse for the amplifier and $2\pi$ pulse for the absorber). So far the relevance of these solutions for present passively mode-locked dye lasers cannot be decided for sure. Assuming the measured spectra of the dye molecules to be throughout homogeneously broadened the pulse intensities measured are too small by about one order of magnitude for the coherent regime to occur. On the other hand, if inhomogeneous broadening plays a part, these coherent effects might come into play already at lower pulse intensities. Once established the coherent interaction could be an additional mechanism to stabilize the pulse regime as well as to overcome the limit of the pulse duration that was given by the finite width of the amplifying and absorbing transition in the 'ordinary' modelocking regime.

## 5.3 The "soliton" laser

In the so-called soliton laser the soliton formation in optical fibers ([5.62], [5.63], see section 4.1) is directly utilized intracavity in combination with an actively modelocked laser producing the seed pulses.

The first laser of this kind was developed in 1983 [5.14] and has been experimentally studied and improved since then [5.15], [5.57]. There are also several theoretical attempts to describe the essentials of this laser [5.58–5.61]. For pulses to evolve in an optical fiber into solitons their wavelength must be in the region of negative GVD and low energy loss. Considering typical single mode glass fibers this means $\lambda > 1.3$ $\mu$m which makes the color center laser [5.64], [5.65] very attractive as pulse source. Figure 5.14 shows the cavity configuration used in the soliton laser. A synchronously pumped color center laser (mirrors $M_1$, $M_2$, $M_0$) is coupled to a retroreflector that contains a beam splitter (S), a polarization preserving single mode fiber, focusing objectives $L_1$, $L_2$ and an end mirror $M_3$. The pulse round trip time in the fiber arm has to be an integral multiple of the one in the main cavity. Moreover, the two coupled cavities have to be matched in length with interferometric accuracy to ensure that the pulses from the fiber arm interfere constructively with those circulating in the

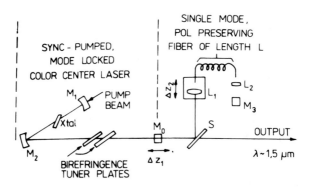

**Figure 5.14** Schematic of the soliton laser (from [5.14]).

main cavity. Accordingly, a control loop to stabilize this phase matching improved the laser performance [5.15]. The color center laser produces pulses at wavelengths around 1.5 $\mu$m, i.e., in the region where for the fiber $\tilde{\kappa}_2 \, k_L'' < 0$ holds and therefore solitons can develop (see section 4.1). While the actively modelocked color center laser alone emits pulses of about 8 ps duration [5.64] pulses as short as 210 fs [5.14] and 60 fs [5.57], respectively, were observed from the soliton laser. This considerable reduction of the pulse duration can be explained by the soliton narrowing, which the pulses experience by passage through a fiber of suitable length (section 4.1). The narrowed pulse is then coupled back into the main cavity where it is amplified and then sent again into the fiber arm. It was experimentally proved that this process proceeds until the pulses in the fiber become solitons, where generally N = 2 solitons were observed and the fiber length L corresponds to half the soliton period $(2L = z_0, (4.29b))$. In particular, the dependence of the pulse duration on the fiber length and the input pulse power shows the characteristic behavior known from the (N = 2) soliton propagation in fibers. From the latter, one would expect that the pulses from the soliton laser shorten when the length of the fiber is decreased. (The periodicity length varies with the square of the pulse duration, see (4.29b)). However, the pulse intensity required in the control fiber which rises inversely proportional with fiber length (4.29a) is limited by the features of the color-center laser. To overcome these limitations, a fiber of lower dispersion but with comparable nonlinearity was used, a so-called "dispersion-flattened" fiber [5.65], which resulted in pulses as short as 60 fs [5.57]. These pulses could be further shortened in an external fiber to 19 fs [5.57] by soliton-narrowing (see 4.1.5.2), which corresponds to about four optical cycles at the laser wavelength (1.5 $\mu$m). This pulse duration is supposed to be limited by the dispersive properties of the fibers used. Experiments with the pulse propagation through a test fiber also revealed that the laser pulses correspond to solitons which require a fiber length to reproduce themselves that exceeds the length of the control fiber considerably, see Fig. 5.15. This means the pulses retroflected in the main cavity are significantly shorter than the pulses launched into the fiber arm. In this sense the fiber cavity acts as an intracavity compressor counterbalancing the spreading mechanisms in the main cavity [5.57].

For a detailed theoretical description of the soliton laser, the complex theory of synchronous pumping (e.g. [5.66–5.71]) would have to be combined with the treatment of pulse propagation through optical fibers. To deal with this complicated mathematical problem, the present theoretical descriptions of the soliton laser make use of simplifications idealizing certain aspects of the laser. An analytical approach was performed in [5.58], it was based on the simplification of a single-cavity laser that

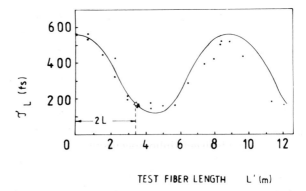

**Figure 5.15** Duration of pulses from test fiber of length L' when sech²-shaped pulses from the soliton laser ($\tau_{L0} = 560$ fs) were launched into it. The solid line is the theoretically expected value for a (N = 2) soliton. The power level (in both control and test fiber) was equal to the (N = 2) soliton power as calculated from the fiber parameters (from [5.15]).

contained both active medium (color center crystal and fiber). Using the round trip model the steady-state regime was explained in the following way.

In the main cavity the pulses are lengthened by limiting bandwidth via the gain medium and the birefringence tuner plates and experience gain. This was taken into consideration in [5.58] by the transfer function

$$J_g = 1 + g\left(e^{-\frac{1}{2}\Delta\omega_m^2 \eta^2} + \frac{1}{\Delta\omega_q^2}\frac{d^2}{d\eta^2}\right) \tag{5.6}$$

where $\Delta\omega_q$ describes the cavity bandwidth, the reciprocal value of $\Delta\omega_m$ is a measure of the temporal gain window determined by the synchronous pump and g is the gain coefficient. The gain has to compensate for the linear losses (mirrors, beam splitter, fiber coupler) and losses associated with the soliton formation. A pulse of a certain energy at the input of the fiber which differs in shape from the soliton develops into a soliton, where the excess energy is temporally dispersed and moves away from the solitary pulse. If the temporal gain window is suitably set, this radiation is not amplified but suppressed after being coupled back into the main cavity. The amplification process and the spectral filtering again lead to a non-soliton pulse which is reshaped into a soliton in the fiber arm. The shaping through the fiber was calculated by means of the inverse scattering method (section 4.1.3) which yields an operator for the fiber arm. The steady state condition was then characterized in terms of the pole locations of the inverse scattering problem. As a result, it became obvious that the main pulse-shaping mechanism is the propagation through the fiber rather than through the main laser cavity containing the synchronously pumped color center sample. The overestimation of the role of the shaping in the fiber resulting from the assumption of a single cavity is probably the reason why this analytical approach fails in the detailed explanation of later experiments [5.72], [5.15] in particular, where the fiber is shorter than the soliton period. Other authors use numerical procedures to simulate the pulse evolution in the soliton laser [5.59–5.61], where for simplicity the pulse shaping in the synchronously pumped color center sample was strongly idealized (inhomogeneously broadened amplifier) but the actual two-cavity arrangement was considered. In [5.61] it could be shown that there are steady states where the pulses do not travel a complete soliton period in the fiber which agrees with the experimental findings [5.57]. However, a number of questions are still open and require further improvements in theory in order to be answered. Despite of the problems in the treatment of the color center laser itself, certain preconditions necessary for applying the NLSE become questionable (e.g. SVEA, neglect of higher-order dispersion, Raman effect) at least when dealing with the theoretical limits of this kind of soliton laser.

As outlined in section 4.3 Raman processes can be utilized to support soliton propagation in fibers at Stokes-shifted wavelengths. Moreover, using pulses of suitable wavelengths, solitons can be amplified and reshaped through SRS which has particular importance in long fibers, where linear losses need to be compensated [5.73–5.75]. On the other hand these effects are used in oscillator schemes — so-called fiber Raman soliton lasers — where the amplification is due to Raman scattering

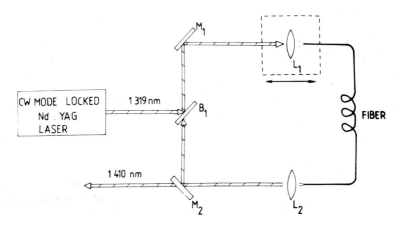

**Figure 5.16**   Scheme of a fiber Raman soliton laser.

stimulated e.g. by the pulse sequence of a modelocked laser [5.76–5.78]. Various oscillator configurations have been developed, one of the simplest arrangements is shown in Fig. 5.16 (e.g. [5.73], [5.77], [5.78]). The pulses from a cw modelocked Nd:YAG laser are coupled through a beam splitter $B_1$ into a ring resonator configuration containing an optical single mode fiber. While travelling through the fiber the pump pulses at 1.319 μm produce Stokes pulses (1.41 μm, see section 4.3) which in turn undergo soliton shaping and are partially coupled out at mirror $M_2$. In the steady state the losses the Stokes pulses experience during one round trip are compensated by SRS induced by the pump pulses.

## 6. MEASUREMENTS OF AND WITH PHASE-MODULATED PULSES

For the application of ultrashort light pulses, it is desirable to know the pulse parameters as exact as possible. On one hand this information is needed for optimum adjustment of the laser itself and subsequent pulse shaping devices (cf. chapters 3, 4, 5). On the other hand pulse chirp can decisively affect the light-matter interaction and, thus, can not be neglected in interpreting the experimental findings in such cases. The latter is particularly important in processes which are influenced or controlled by phase modulation, e.g., pulse compression, (coherent) wave mixing. Also, if the pulses are to be applied in time-resolved measurements, the resolution limit can be considerably lower than the pulse duration in making use of suitable deconvolution procedures. This, however, requires accurate knowledge about the pulse shape, duration and phase modulation. Similarly, an optimal compression of chirped pulses requires information about the pulse chirp $\left( \dfrac{d}{dt} \Phi(t) \right)$ and the corresponding spectral phase $\phi(\omega)$ for the choice of a suitable linear element or a combination of them. This is especially desirable where it is necessary to fit contributions of $\phi(\omega)$ higher than second order which, e.g., was the case with the compression of light pulses to 6 fs [6.1] (cf. chapter 4).

It is also worthwhile to mention that new measuring techniques are being developed to utilize chirped light pulses for measuring and exciting various linear and nonlinear processes.

### 6.1 Determination of pulse duration and chirp

*Correlation techniques.* The pulse duration has been accurately measured with intensity correlation techniques using nonlinear optical processes (e.g. [6.2]). From the autocorrelation of second order

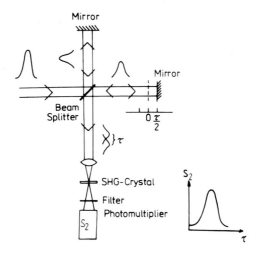

**Figure 6.1** Scheme of an optical correlator based on second harmonic generation (SHG). With the collinear beam geometry autocorrelation with and without background can be measured using parallel or perpendicular polarization of the partial pulses and the corresponding type of phase matching.

$$S_2(\tau) \propto \int\limits_{-\infty}^{\infty} dt\ I(t)\ I(t-\tau) \qquad (6.1)$$

the pulse duration can be evaluated if a certain pulse shape is assumed. The pulse duration $\tau_L$ is related to the FWHM of the autocorrelation function $\tau_{AC}$ via the relation

$$\tau_{AC} = c_0\, \tau_L \qquad (6.2)$$

where $c_0$ is a number that depends on the actual pulse shape taking on for example values of 1.41, 1.54 and 2 for Gaussian, sech — and Lorentzian-shaped pulses, respectively (for a comprehensive listing of $c_0$ for various pulse shapes see e.g. [6.2]).

Figure 6.1 shows an experimental set up for measuring the autocorrelation by means of second harmonic generation (SHG). By choosing a suitable beam geometry and type of phase matching the correlation can be measured with or without background (6.1).

Concerning fs light pulses certain aspects of the second harmonic generation have to be checked carefully for the correlation function to yield a measure of the pulse duration. In particular, the conversion bandwidth must be large enough to ensure that all spectral components of the pulses to be measured contribute to the second harmonic signal, i.e. the conversion process must not be wavelength dependent. If this requirement is fulfilled and the wavelength scales of fundamental and second harmonic radiation are properly adjusted, the square root of the second harmonic spectrum fits the spectrum of the initial pulse. A narrower second harmonic spectrum would indicate bandwidth limitation in the conversion process which may lead to an autocorrelation function which counterfeits broader input pulses.

Since the nonlinear susceptibilities can be assumed as frequency-independent within frequency intervals $\Delta\omega$ typical for fs light pulses ($\Delta\omega \leqslant 10^{-14}$ s) the requirement of a constant conversion efficiency provides first of all a condition for the product of the phase mismatch and the sample length $\Delta k(\omega)L$. Assuming $\Delta k = 0$ at the center frequency $\omega_L$ it is sufficient for a rough estimation to require that the phase mismatch at the output of the sample is less than $\pi/2$ over the entire pulse spectrum, i.e.

$$\alpha = \left| \frac{d}{d\omega} \Delta k \right| \Delta\omega L < \frac{\pi}{2} \qquad (6.3)$$

where $\Delta k(\omega)$ has to be calculated for the actual type of phase matching and is directly related to the group velocity dispersion. For example, in type I phase matching (oo-e) in negatively birefringent crystals $d\Delta k/d\omega$ is given by

$$\frac{d}{d\omega} \Delta k = \frac{d}{d\omega} [k_{2\omega}^e(2\omega) - 2k_\omega^0(\omega)] \qquad (6.4)$$

The possible sample length L varies with the type of the crystal and the kind of phase matching and is, e.g., about 100 $\mu$m (Urea) and about 1 mm (KDP) if 50 fs pulses are to be measured. Instead of changing the sample length, often the confocal parameter of the focused light is adjusted by appropriate (tight) focussing. A detailed theoretical investigation of the correlation measurements with various types of phase matching and crystal parameters is found in [6.3]. As shown there the distortion of the intensity correlation for $\alpha \neq 0$ is dependent on the actual pulse shape and is particularly pronounced if the pulse to be measured is chirped. With (6.3) being valid, the correlation (6.1) is always symmetrical with respect to $\tau = 0$ and rather insensitive to the actual pulse shape.

In principle, higher order correlation functions of the type

$$S_{n+1}(\tau) \propto \int_{-\infty}^{\infty} dt \ I^n(t) \ I(t-\tau) \qquad (6.5)$$

contain detailed information on the pulse shape since $I^n(t)$ can be considered as a sampling pulse for $I(t - \tau)$. However, in order to generate $I^n(t)$ higher order nonlinear processes have to be applied which require sufficiently high pulse intensities. This together with the large conversion bandwidth needed and the high experimental effort limit the use of such techniques at least for unamplified femtosecond pulses.

The correlation functions given in (6.1) and (6.5) are pure intensity correlations and hence they contain no phase information of the signal. This disadvantage can be overcome by changing the delay in the correlator (Fig. 6.1) with interferometric accuracy. Instead of the second order intensity autocorrelation (with backround) now the so-called interferometric autocorrelation (IAC) is measured [6.4]

$$S_I(\tau) \propto \int_{-\infty}^{\infty} dt \, | \, [E_1(t) + E_2(t - \tau)]^2 \, |^2 \qquad (6.6)$$

Replacing $E_{1,2}(t) = A_{1,2}(t)e^{i\Phi_{1,2}(t)} e^{i\omega_L t} + $ c.c. and averaging over the fast time oscillations which is automatically done in the detection system (6.6) can be rewritten as

$$S_I(\tau) \propto \int_{-\infty}^{\infty} A_1^4(t)dt + 4 \int_{-\infty}^{\infty} A_1^2(t)A_2^2(t - \tau)dt + \int_{-\infty}^{\infty} A_2^4(t)dt$$

$$+ 2 \int_{-\infty}^{\infty} A_1^2(t)A_2^2(t - \tau) \cos[2(\omega_L\tau + \Phi_1(t) - \Phi_2(t - \tau)]dt \qquad (6.7)$$

$$+ 4 \int_{-\infty}^{\infty} \cos[\omega_L\tau + \Phi_1(t) - \Phi_2(t - \tau)](A_1^3(t)A_2(t - \tau) + A_1(t)A_2^3(t - \tau))dt$$

The underlined terms correspond to the background free intensity autocorrelation (6.1) (dashed line) and to the intensity correlation with background (solid line), respectively, which are usually measured. It is obvious that $S_I(\tau)$ is sensitive to a time dependent phase $\Phi_{1,2}(t)$ in a rather complicated manner. Roughly speaking, the width of the IAC decreases with increasing phase modulation (smaller coherence length) while the intensity autocorrelation preserves its shape (Fig. 6.2).

A straightforward calculation of $\dot\Phi(t)$ from the IAC is only feasible if pulse shape and functional dependence of $\dot\Phi(t)$ are known. For example, the chirp parameter of a linearly chirped pulse of known shape can be

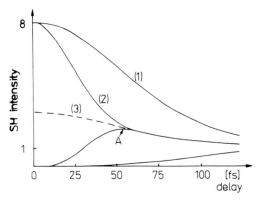

**Figure 6.2** Envelope of the interferometric autocorrelation of a sech$^2$ pulse of 80 fs duration without (1) and with chirp (2). For comparison the intensity autocorrelation (3) is shown. The maximum of the lower envelope is denoted by ''A''.

evaluated from the locus of the maximum (A) of the lower envelope (Fig. 6.2).

In the general case, the intensity profile and the time dependent phase can be determined if additional information is available. This information can be gained from the intensity autocorrelation, the pulse spectrum [6.5] and the intensity and the interferometric crosscorrelation of the partial pulse $I_1$ with the other partial pulse $I_2$ after having traversed a suitable linear element with a well defined optical transfer function (e.g., an element with GVD). A simultaneous fitting was performed in [6.6] which yielded the desired information on pulse shape and phase behavior (Fig. 6.3) of the pulses from a fs dye laser. Despite the difficulties arising from the numerical fitting procedures, this is the only method that is applicable for the extremely short fs light pulses so far.

*Self-heterodyning.* Another technique — the so-called self-heterodyning — was demonstrated in [6.7] to determine modulus and phase of a pulse shaped in an optical fiber. The pulses were produced in a synchronously pumped dye laser and were tuned to be in resonance with one of the sodium lines. In an optical fiber they become broader and almost linearly chirped (see section 4.1). The device for measuring this phase modulation essentially consists of a correlator (Fig. 6.1), where however, the partial pulses are modified in each of the two arms in a well defined manner. In the first arm the chirped pulses travel through a grating compressor where the compressed pulses then act as δ-sampling pulses for the original pulses that pass through the second arm. So in a first

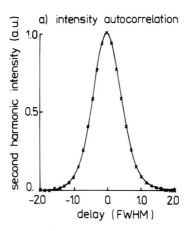

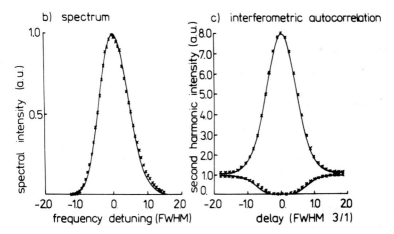

**Figure 6.3** Example of pulse shape determination through fitting of a. intensity auto-correlation (measured without background), b. pulse spectrum, c. interferometric auto-correlation. The crosses are the experimental data points. The electric field was found to vary as

$$\overline{E}(t) = e^{-0.15i(t/\tau)^2} \frac{1}{\exp\left(-\dfrac{t}{0.75\tau}\right) + \exp\left(\dfrac{t}{1.25\tau}\right)}$$

where $\tau_L = 76$ fs ($\simeq 1.72\tau$) (from [6.5]).

step the crosscorrelation yields the pulse shape occurring at the output of the fiber (Fig. 6.4a). In a second step, a sodium vapor cell is inserted in the second arm. The interaction of the pulses with the resonant atomic systems leads to a fast modulation of the pulse intensity. This modulation

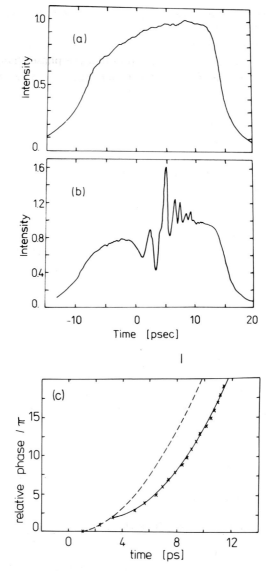

**Figure 6.4** Example for measuring the modulus and phase of a pulse chirped and broadened in an optical fiber, a. intensity crosscorrelation of the pulse to be measured and a much shorter pulse obtained by compression in a grating compressor placed in one arm of the correlator, b. intensity crosscorrelation between the pulse after interaction with a resonant system (sodium vapor) and the compressed pulse, c. phase as a function of the time. The phase curve shows an almost parabolic behavior indicating a linear frequency modulation behind the fiber. The dashed curve is the phase expected from a linearly chirped pulse relative to the other sodium line ($D_1$) (from [6.7]).

can be time-resolved by measuring the crosscorrelation with the compressed pulse (Fig. 6.4b).

$\Phi(t)$ can now easily be evaluated apart from an unimportant constant phase $\Phi_0$ when taking into consideration that the phase change between two successive maxima of the curve shown in Fig. 6.4b is $2\pi$. Accordingly the curve $\Phi(t)$ is obtained by counting the maxima and minima (i.e. the phase changes by $\pi$) as a function of the time t (Fig. 6.4c).

The physical mechanism that shapes the pulses in this favorable manner is associated with the propagation of small-area pulses which had importance in the explanation of transient phenomena in spin resonance [6.8], [6.9] and was then theoretically described in [6.10] and experimentally found in [6.11] for the optical case. In terms of our notation (see section 4.2) the pulse propagation through the weakly absorbing atomic vapor can be briefly described as follows.

Due to the narrowness of the absorbing transition (the transition width is much smaller than the pulse spectrum) and the relatively weak electric field, the occupation numbers of the (two-level) atomic system can be treated as constant.

From (4.45) and (4.46) the polarization in the frequency domain can be written as

$$\underline{P}(\omega,z) = -\frac{i}{\hbar}\, q\, \frac{S_1\, \mu_{12}}{\dfrac{1}{T_2} + i(\omega - \omega_{21})}\; \underline{\overline{E}}(\omega - \omega_L)\, e^{-i(\omega_L t - k_L z)} + \text{c.c.} \qquad (6.8)$$

where we omitted the index NL since, because of $S_1 \neq S_1(|E|^2)$, $\underline{P}(\omega,z)$ can be regarded as a linear polarization and is equivalent to

$$\underline{P}(\omega,z) = \epsilon_0[\epsilon(\omega) - 1]\, \underline{E}(\omega,z) \qquad (6.9)$$

(cf. equation (2.14)).

If we compare (6.8), (6.9), the general dispersion relation (2.15c) yields for the propagation constant $k(\omega)$

$$k(\omega) = \frac{\omega}{c}\left[1 - i\, \frac{\mu_0\, c^2 q_1 |\mu_{12}|^2 T_2}{2\hbar} \cdot \frac{1}{1 + i(\omega - \omega_{21})T_2}\right] \qquad (6.10)$$

where we made use of the fact that the second term in (6.10) is much smaller than 1.

With (6.10) and (2.15b) we can build up the solution of the wave equation which is

$$\underline{E}(\omega,z) = \underline{E}(\omega,0) e^{-i\frac{\omega}{c}z}\, \exp\left\{-\frac{1}{2}\, \sigma_{12} q_1 z\, \frac{1}{1 + i(\omega - \omega_{21})T_2}\right\} \qquad (6.11)$$

where $\sigma_{12}$ is defined according to (4.50c) as the absorption cross section in the transition center.

Assuming optically thin samples ($\sigma_{12}q_1z \ll 1$) the second exponential function in (6.11) is expanded up to first order. If we now introduce the pulse envelope (see (2.4)) and carry out the back transformation of the modified equation (6.11) into the time domain we find

$$A(t,z) = A(t,0) - \frac{q_1\sigma_{12}z}{2 T_2} e^{-t/T_z} \int_{-\infty}^{t} A(t',0)e^{t'/T_2} dt' \qquad (6.12)$$

The intensity measured at the output of the sample is proportional to $|A(t,z)|^2$ and can thus be considered as the superposition of the incident field and a field arising from the response of the sample [6.7]. The atomic system is excited to produce polarization oscillations at the frequency $\omega_{21}(=\omega_L)$. These oscillations interfere with the incident electric field and produce in analogy to an optical heterodyne modulations at the difference frequency that contain the information on the phase modulation. ($\omega(t) - \omega_L = \dot{\Phi}(t)$).

Indeed, a closer inspection of (6.12) shows the occurrence of these oscillations. For the simplest incident wave of constant amplitude and linear frequency chirp

$$A(t,0) = A_0 e^{ibt^2} \qquad (6.13)$$

the output signal $|A(t,z)|^2$ for the limiting case $b \to \infty$ is given by [6.7]

$$|A(t,z)|^2 = A_0^2 + \left(\frac{q_1\sigma_{12}z}{2 T_2}\right)^2 \frac{2\pi}{b} u^2(t)e^{-2t/T_2}$$
$$- 2A_0 \left(\frac{q_1\sigma_{12}z}{2 T_2}\right) \sqrt{\frac{2\pi}{b}} \underline{u(t) \cos\left(bt^2 + \frac{\pi}{4}\right)} \qquad (6.14)$$

where u(t) is the unit step function. Obviously the underlined term in (6.14) is responsible for an intensity modulation, characterizing the time dependent phase $\Phi(t) = bt^2$.

The main disadvantages of the technique described above are the need for a sampling pulse which is shorter than the pulse to be measured and the exact tuning of the laser wavelength to the center of the atomic transition. Because equation (6.11) is completely analogous to (2.21), which describes an arbitrary linear passive element, the second restriction could be principally overcome by using, for example, interferometric structures or masking in the frequency spectrum instead of resonant atoms. Such a technique was applied in [6.12a] which together with some improvements in the evaluation revealed a nonlinear behavior of the chirp at the pulse edges.

The need for shorter pulses could be avoided by additional broadening of the pulse to be characterized by linear optical samples in a well defined manner before recording the crosscorrelation with the initial pulse [6.12b] and the determination of the phase $\Phi(t)$. The time dependent phase of the initial pulse can be obtained after deconvolution of the known optical transfer function of the linear element used for the broadening.

## 6.2  Experiments with light pulses using the shaping in modulus and phase

In the section above we described techniques where the pulse characteristics were changed in a manner that allows us to conclude about the properties of the original pulse. On the other hand, if the pulse parameters are known, one can determine certain optical properties of elements which are traversed from the change that the pulse experiences. This was demonstrated in measuring the phase and amplitude response of dielectric mirrors [6.13].

The corresponding experimental set up consists of a correlator similar

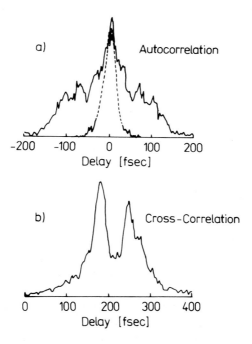

**Figure 6.5**  Pulse distortion through reflection from a broadband dielectric mirror, a. autocorrelation before (dashed line) and after reflection (solid line) b. crosscorrelation between the original pulse and the pulse after reflection from the mirror (from [6.13]).

to that shown in Fig. 6.1 where both end reflectors are assumed not to distort the pulse. After the intensity autocorrelation has been recorded (Fig. 6.5a, dashed line) one of the end mirrors is replaced by the mirror to be measured. The crosscorrelation (Fig. 6.5b) indicates a substructure of the reflected pulse which could be attributed exclusively to the phase response of the mirror (section 3) because of its constant high reflectivity over the spectral range of interest.

While this extracavity technique requires relatively strong effects in order to resolve the pulse characteristics by comparing intensity auto-correlation and crosscorrelation, intracavity measurements [6.14], [6.15] provide a more sensitive technique. The high sensitivity rests on the multiple pass of the pulses through the resonator elements, which enhances the effect induced by a small perturbation of one of the cavity parameters. Moreover, the pulse characteristics are very sensitive to changes of some parameters of optical elements in the cavity. If we insert a sample in the resonator the modulus and phase of the output pulse is changed in a characteristic manner. In general, however, it is impossible to calculate the optical properties of the sample from the (measured) modi-fication of the pulses. As an alternative a "zero method" was used [6.14, 6.15], where the pulse properties of the laser are restored by modifying a "calibrated" optical intracavity component to compensate for the

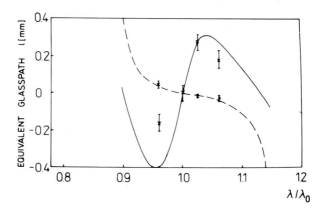

**Figure 6.6** Measurement of GVD of two sets of dielectric, single-stack multilayer mirrors with a compensation method based on the adjustment of the intracavity glass path. For the measurements one of the mirrors (reference mirror) of a CPM ring laser was replaced by the mirror to be characterized which differs in its resonance wavelength $\lambda$. Then the intracavity glasspath was readjusted to the same operating regime as obtained with the reference mirror ($\lambda = \lambda_0$). The lines correspond to calculations according to (6.15), where $\psi^{(2)}(\omega)$ was calculated by means of the usual matrix formalism (see section 3) and using the parameters of Fig. 3.9 (from [6.14]).

"perturbation" resulting from the sample. This compensation method is particularly sensitive if one adjusts the "zero point" to be at the end of the stability range of the laser, where small deviations of the laser parameter have drastic effects.

This technique was used to determine the dispersive properties of dielectric multilayer mirrors by counterbalancing their dispersive effect with the intracavity glass path l in a fs dye laser (Fig. 6.6). The relation between l and the GVD introduced by the mirror is

$$\frac{d^2\psi}{d\omega^2} = \frac{l}{c}\left[2\,\frac{dn}{d\omega} + \omega\,\frac{d^2n}{d\omega^2}\right] \tag{6.15}$$

where n is the refractive index of the glass. In principle, higher order dispersion can also be measured if one finds suitable linear elements (or a combination of them) through which it is possible to adjust the different dispersion orders independently.

As already mentioned, phase modulation of light pulses is particularly important in experiments where phase coherence plays a role. A typical process is four wave mixing (FWM) which has been successfully applied on a fs time scale for the measurement of phase relaxation transients [6.16], [6.17] and for pulse compression [6.18]. It should be noted that the pulse compression observed in [6.18] is simply a result of the fact that the FWM is a nonlinear process of third order and does not necessarily require phase modulation.

The influence phase modulation can have on FWM will be described on the basis of the experimental determination of phase relaxation times $T_2$ in organic dyes [6.16]. A suitable experimental arrangement is sketched in Fig. 6.7a. Two counterpropagating pump pulses (1,2) meet exactly in an absorbing dye and interact with a delayed (weaker) probe pulse having the same wavelength (degenerate four wave mixing — DFWM). This beam geometry as well as the sufficiently high intensity of the pump pulses were assured by using the saturable absorber of a CPM laser as sample, where one of the output pulses served as probe pulse. Each of the pulses excites a polarization oscillation which subsides for a time $T_2$. The DFWM can be regarded as an interference effect of the corresponding polarization waves propagating in different directions, where the result is the formation of gratings in the overlapping range. The DWFM signal, which propagates collinearly but opposite to the probe pulse can then be explained as the result of a partial reflection of pulse 1 at the grating which was formed by the polarization waves belonging to the probe pulse and pulse 2.

The DWFM signal measured as a function of the delay of the probe pulse contains the information about the dephasing times via the life times of the induced gratings. If pulse shape and chirp are exactly known the $T_2$

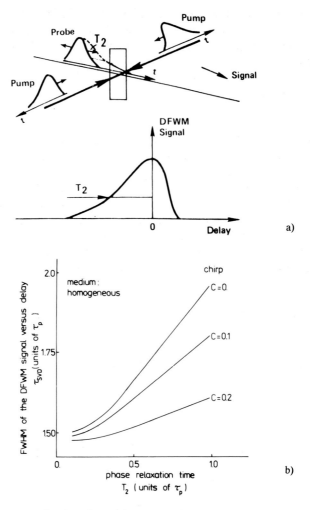

**Figure 6.7** Influence of pulse chirp on four-wave mixing experiments. a. Geometry of degenerate four-wave mixing (DFWM). The signal is measured as a function of the delay that the proble pulse has with respect to the colliding pump pulses. b. Calculated FWHM of the DFWM signal as a function of the phase relaxation time for different chirp parameters of the pulses. The electric field of the probe and pump pulses is proportional to

$$\overline{E}(t) \propto e^{-ic(t/\tau)^2} \operatorname{sech}(t/\tau)$$

(from [6.16]).

time can be determined from the width of the DFWM signal $\tau_{SVD}$ versus delay after deconvolution. The effect of phase modulation upon $\tau_{SVD}$ is illustrated in Fig. 6.7b where $\tau_{SVD}$ versus $T_2$ is shown for different chirp parameters assuming linearly chirped sech-shaped pulses. Obviously even small chirp parameters would lead to a sizable misinterpretation of the experimental results.

The drastic influence that pulse chirp can have even on the results of time-resolved absorption recovery was investigated in [6.19], where the transient absorption recovery in GaAs was measured, see Fig. 6.8. In experiments of this kind the chirp effect is particularly pronounced if the processes to be measured proceed on a time scale which is comparable with the pulse duration and time intervals in which the first derivative of the pulse phase (i.e. instantaneous frequency) changes on the order of the magnitude of the inverse pulse duration. The different relaxation characteristics shown in Fig. 6.8 are due to a rapid intraband relaxation, i.e., the occurrence of carriers of decreasing energy at later times (corresponding to lower excitation frequencies) which means that the instantaneous pulse frequency, and with it the excitation, follows the relaxation process if the chirp is negative or has the opposite direction if it is positive. This obviously would result in strongly differing relaxation parameters if only the pulse duration was considered in the deconvolution of the curves measured.

Finally we want to mention a method to shape frequency swept light

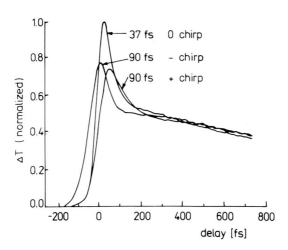

**Figure 6.8** Absorption saturation dynamics of GaAs with fs pulses of equal bandwidth but different chirp (from [6.19]).

pulses which gained importance with respect to the extracavity pulse compression (section 4.1). We have already pointed out that in some cases the compressed pulse exhibits wings or subpulses that often result from undesired frequency components of the pulses chirped in the fiber. These frequency components occur most likely at the pulse edges and indicate a deviation from the linear chirp. In [6.20], [6.21], [6.22] it was demonstrated that these frequencies can be cut off by simple geometric masking of the frequency spectrum within the grating compressor where the different frequency components are distributed across the beam cross section. This leads to an improvement in the pulse quality although it is often accompanied by a slight pulse broadening.

### Acknowledgement

The authors are sincerely grateful to Dr. V. Petrov for numerous valuable discussions and wish to thank Dipl. Phys. G. Werner and Dipl. Phys. P. Heist for the critical reading of the manuscript. The authors also want to thank J. Harnisch and K. Truckenbrodt for assistance in the technical preparation of the manuscript.

### REFERENCES

1.1     S.L. Shapiro (Ed.): Ultrashort Light Pulses, Springer, New York (1977)
1.2     J. Herrmann, B. Wilhelmi: Lasers for Ultrashort Light Pulses, Akademie Verlag, Berlin (1986)
1.3     H.W. Mocker, R.J. Collins: *Appl. Phys. Lett.* **7** (1965) 270
1.4     A.J. DeMaria, D.A. Stetser, H. Meynan: *Appl. Phys. Lett.* **8** (1966) 174
1.5     M.B. Ketchen, *et al.*: *Appl. Phys. Lett.* **48** (1986) 751
1.6     R.L. Fork, B.I. Greene, C.V. Shank: *Appl. Phys. Lett.* **38** (1981) 671
1.7     I.A. Valdmanis, R.L. Fork, J.P. Gordon: *Opt. Lett.* **10** (1985) 131
1.8     W. Dietel. J.I. Fontaine, J.C. Diels: *Opt. Lett.* **8** (1983) 4
1.9     J.C. Diels, J.J. Fontaine, I.C. McMichael, B. Wilhelmi, W. Dietel, D. Kühlke, W. Rudolph: *Kvant. Electron.* **10** (1983) 2398 (*Sov. J. Quant. Electr.* **13** (1983) 1562)
1.10    O.E. Martinez, J.P. Gordon, R.L. Fork: *J. Opt. Soc. Am.* **A1** (1984) 1003
1.11    J. Kuhl, J. Heppner: *IEEE J. Quant. Electr.* **QE22** (1986) 182
1.12    M. Yamashita, M. Ishikawa, K. Torizuka, T. Sato: *Opt. Lett.* **11** (1986) 504
1.13a   W.H. Knox, R.L. Fork, M.C. Downer, R.H. Stolen, C.V. Shank, J.V. Valdmanis: *Appl. Phys. Lett.* **46** (1985) 1120
     b  R.L. Fork, C.H. Brito-Cruz, P.C. Becker, C.V. Shank: *Opt. Lett.* **12** (1987) 483
1.14    Technical Digest of Spectra Physics (1986)
1.15    E.B. Treacy: *IEEE J. Quant. Electr.* **QE5** (1969) 454
1.16    T.K. Gustafson, J.P. Taran, H.A. Haus, J.R. Lifshitz, P.L. Kelley: *Phys. Rev.* **177** (1969) 539
1.17    A. Laubereau: *Phys. Lett.* **29A** (1969) 539
1.18    M.A. Duguay, J.W. Hansen: *Appl. Phys. Lett.* **14** (1969) 14
1.19    C.V. Shank, R.L. Fork, R. Yen, R.H. Stolen, W.J. Tomlinson: *Appl. Phys. Lett.* **40** (1982) 761

1.20 D. Grischkowsky, A.S. Balant: *Appl. Phys. Lett.* **41** (1982) 1
1.21 B. Nikolaus, D. Grischkowsky: *Appl. Phys. Lett.* **42** (1983) 1
1.22 B. Nikolaus, D. Grischkowsky: *Appl. Phys. Lett.* **43** (1983) 228
1.23 J.G. Fujimoto, A.M. Weiner, E.P. Ippen: *Appl. Phys. Lett.* **44** (1984) 832
1.24 R. Meinel: *Opt. Comm.* **47** (1983) 343
1.25 R.L. Fork, C.V. Shank, R.T. Yen, C.A. Hirliman: *IEEE J. Quant. Electr.* **QE19** (1983) 500
1.26 B. Wilhelmi: *Ann. Phys.* **43** (1986) 355
1.27 T. Damm, M. Kaschke, F. Noack, B. Wilhelmi: *Opt. Lett.* **10** (1985) 176
1.28 K. Tai, A. Tomita: *Appl. Phys. Lett.* **48** (1986) 1033
1.29 E.M. Dianov, A.J. Karasik, P.W. Mamyshev: *Kvant. Elektron.* **11** (1984) 1078
1.30 S.A. Akhmanov, W.A. Vyslouch, A.S. Chirkin: *Usp. Fiz. Nauk.* **149** (1986) 450
1.31 V.I. Karpman: *Nonlinear Waves in Dispersive Systems*, Nauka, Moscow (1973)
1.32 D.H. Auston, K.B. Eisenthal (Eds.): *Ultrafast Phenomena IV*, Springer, Berlin, Heidelberg, New York, Tokyo (1984)
1.33 G.R. Fleming, A.E. Siegman (Eds.): *Ultrafast Phenomena V*, Springer, Berlin, Heidelberg, New York, Paris, Tokyo (1986)
1.34 Proceedings of "Ultrafast Phenomena in Spectroscopy UPS'83'', Minsk (1983)
1.35 E. Klose, B. Wilhelmi (Eds.): Proceedings of "Ultrafast Phenomena in Spectroscopy UPS'85'' Teubner, Leipzig (1986)
1.36 N. Nakano, H. Kumoda: *Phys. Rev.* **35** (1987) 4712 and references therein
1.37 J.D. Kafka, T. Sizer, I.N. Duling, C.W. Gabel, E. Mourou: in *Picosecond Phenomena III* p. 107, Springer, New York (1982)
1.38 R. B. Marcus, A.M. Weiner, J.H. Abeles, P.S.D. Lin: *Appl. Phys. Lett.* **46** (1986) 357

2.1 M. Born, E. Wolf: Principles of Optics, Pergamon, Oxford, New York, Toronto, Sydney, Frankfurt (1980)
2.2 J.D. Jackson: Classical Electrodynamics, Wiley, New York (1975)
2.3 Y.R. Shen: The Principles of Nonlinear Optics, Wiley, New York (1985)
2.4 M. Schubert, B. Wilhelmi: Nonlinear Optics and Quantum Electronics, Wiley, New York (1986)
2.5 see [1.17]
2.6 W.J. Tomlinson, R.H. Stolen, C.V. Shank: *J. Opt. Soc. Am.* **B1** (1984) 139
2.7 P. Sperber, A. Penzkofer: *Opt. Commun.* **54** (1985) 160
2.8 H. Nuthel, D.M. Guthals, J.H.I. Clark: in Technical Digest of CLEO'84 p. 182, Anaheim (1984)

3.1a L.D. Landau, E.M. Lifshitz: Course of Theoretical Physics, Addison, New York (1961)
   b Optics Guide 3, Melles Griot (1985)
3.2 see [1.9]
3.3 D.H. Auston, K.P. Cheung: J. Opt. Soc. Am. B2 (1985) 606
3.4 see [1.15]
3.5a see [1.10]
   b R.L. Fork, O.E. Martinez, J.P. Gordon: *Opt. Lett.* **9** (1984) 150
3.6 C. Froehly, B. Colombeau, M. Vampouille: in Progress in Optics, Vol. XX p. 115, North-Holland (1981)
3.7a O.E. Martinez: *J. Opt. Soc. Am.* **B3** (1986) 929
   b O.E. Martinez: *IEEE J. Quant. Electr.* **QE23** (1987) 59
3.8 Z. Bor, B. Racz: *Opt. Comm.* **54** (1985) 165
3.9 see [2.1]
3.10 V. Petrov, F. Noack, W. Rudolph, C. Rempel: *36 (1988) 167 Exp. Techn. Phys.*
3.11 W. Dietel, E. Döpel, K. Hehl. W. Rudolph, E. Schmidt: *Opt. Comm.* **50** (1984) 179
3.12a S. DeSilvestri, P. Laporta, O. Svelto: *IEEE J. Quant. Electr.* **20** (1984) 533
   b P. Laporta, V. Magni: *Appl. Opt.* **24** (1985) 2014

3.13    A.M. Weiner, J.G. Fujimoto, E.P. Ippen: *Opt. Lett.* **10** (1985) 71
3.14    D.N. Christodoulides, E. Bourkoff, R.I. Joseph, T. Simos: *IEEE J. Quant. Electr.* **QE22** (1986) 186
3.15    F. Gires, P. Tournois: *C.R. Acad. Sc. Paris* **258** (1964) 6112
3.16    J. Desbois, F. Gires, P. Tournois: *IEEE J. Quant. Electr.* **9** (1973) 213
3.17a   J. Heppner, J. Kuhl: *Appl. Phys. Lett.* **47** (1985) 453
    b   J. Kuhl, J. Heppner: *IEEE J. Quant. Electr.* **QE22** (1986) 182
3.18    P.M.W. French, G.F. Chen, W. Sibbett: *Opt. Comm.* **57** (1986) 263
3.19    M. Yamashita, M. Ishikawa, K. Torizuka, T. Sato: *Opt. Lett.* **11** (1986) 504
3.20    M.A. Duguay, J.W. Hansen: *Appl. Phys. Lett.* **14** (1969) 14

4.1     T.Y. Chen: *Opt. Engin.* **20** (1981) 220
4.2     O. Svelto: in Progress in Optics, Vol. XII p. 1, North-Holland, Amsterdam, Oxford, Tokyo, New York (1974)
4.3     H.M. Gibbs, S.L. McCall, T.N.C. Venkatesan: *Phys. Rev. Lett.* **36** (1976) 1135
4.4     R.G. Brewer, C.H. Lee: *Phys. Rev. Lett.* **21** (1968) 267
4.5     A.P. Veduta, B.P. Kirsanov: *Sov. Phys. JETP* **27** (1971) 736
4.6     R.R. Alfano, S.L. Shapiro: *Phys. Rev. Lett.* **24** (1970) 592
4.7     H. Nakatsuka, D. Grischkowsky, A.C. Balant: *Phys. Rev. Lett.* **47** (1981) 910
4.8     see [1.17]
4.9     V.I. Karpman: *Zh. Eksp. Teor. Fiz.* **6** (1967) 829
4.10    A. Hasegawa, F. Tappert: *Appl. Phys. Lett.* **23** (1973) 142
4.11    M. Jain, N. Tzoar: *J. Appl. Phys.* **49** (1978) 4649
4.12    L.A. Ostrovski: *Zh. Eksp. Teor. Fiz.* **51** (1966) 1189
4.13    R.J. Joenk, R. Landauer: *Phys. Rev. Lett.* **A24** (1967) 228
4.14    Y. Kodama: *J. Phys. Soc. Jap.* **45** (1978) 311
4.15    T. Yajima: *Japan. J. Appl. Phys.* **21** (1982) 1044
4.16    P.K.A. Wai, C.R. Menynk, Y.C. Lee, H.H. Chen: *Opt. Lett.* **11** (1986) 464
4.17    see [1.30]
4.18    R.H. Stolen, Ch. Lin: *Phys. Rev.* **A17** (1978) 1448
4.19    F.P. Kapron, D.B. Keck: *Appl. Opt.* **10** (1971) 1519
4.20    D. Marcuse (Ed.): Integrated Optics, *IEEE press*, New York (1973)
4.21    L.G. Cohen, W.L. Mammel: *Electron. Lett.* **18** (1982) 1023
4.22    A.D. Pearson, L.G. Cohen, W.A. Reed, J.T. Krause, E.A. Sigety, F.W. DiMarcello, A.G. Richardson: *IEEE J. Lightwave Technol.* **LT-2** (1984) 346
4.23    F. Calogero, A. Degasperis: Spectral Transform and Solitons, North-Holland, Amsterdam, New York, Oxford (1982)
4.24    V.E. Zakharov, A.B. Shabat: *Zh. Eksp. Teor. Fiz.* **61** (1971) 118 (*Sov. Phys. JETP* **34** (1972) 62)
4.25    see [1.24]
4.26    R.K. Bullough, P.J. Caudrey: Solitons, Springer, Berlin, Heidelberg, New York (1980)
4.27    G.L. Lamb, Jr.: Elements of Soliton Theory, Wiley, New York (1980)
4.28    R.A. Fisher, P.L. Kelley, T.K. Gustafson: Appl. Phys. Lett. 14 (1969) 140
4.29    see [1.20]
4.30    A. Hasegawa, Y. Kodama: *Proc. IEEE* **69** (1981) 1145
4.31    W.J. Tomlinson, R.H. Stolen, C.V. Shank: *J. Opt. Soc. Am.* **B1** (1984) 139
4.32    see [1.26]
4.33    J.D. Kafka, T. Baer: *Opt. Lett.* **12** (1987) 401
4.34    see [1.27]
4.35    V.A. Vislouch, T.A. Matveev: Preprint MGU Fiz. Fak. No. 24, Moscow, (1985)
4.36    E. Bourkoff, W. Zhao, R.I. Joseph, D.N. Christodoulides: *Opt. Lett.* **12** (1987) 272
4.37    see [1.22]
4.38    E.M. Dianov, A.Y. Karasik, P.W. Mamyshev: *Kvantov. Elektr.* **11** (1984) 1078
4.39    J.D. Kafka, B.H. Kolner, T. Baer, D.M. Bloom: *Opt. Lett.* **9** (1984) 505
4.40    A.M. Johnson, R.H. Stolen, W.M. Simpson: *Appl. Phys. Lett.* **44** (1984) 729

4.41   A.S.L. Gomes, U. Österberg, W. Sibett, J.R. Taylor: *Opt. Comm.* **54** (1985) 377
4.42   see [1.28]
4.43   B. Zysset, W. Hodel, P. Beaud, H.P. Weber: *Opt. Lett.* **11** (1986) 156
4.44   B. Valk, K. Wilhelmsson, M.M. Salour: *Appl. Phys. Lett.* **50** (1987) 656
4.45   see [1.2]
4.46   see [1.23]
4.47   J.M. Halbout, D. Grischkowsky: *Appl. Phys. Lett.* **45** (1984) 1281
4.48   see [1.13]
4.49   W. Zhao, E. Bourkoff: *Appl. Phys. Lett.* **50** (1987) 1304
4.50   C. Rolland, P.B. Corkum: *J. Opt. Soc. Am.* **B3** (1986) 1625
4.51a  V. Petrov, W. Rudolph, B. Wilhelmi: CLEO'88, Anaheim (1988) J. Mod. Opt. . . .
       (1989) . . .
    b  C. Rolland, P.B. Corkum: J. Opt. Soc. Am. 35 (1988) 641
4.52   J. Satsuma, N. Yajima: *Progr. Theor. Phys. Suppl.* **55** (1974) 284
4.53   L.F. Mollenauer, R.H. Stolen, J.P. Gordon: *Phys. Rev. Lett.* **15** (1980) 1095
4.54   D. Yervick, B. Hermansson: *Opt. Comm.* **47** (1983) 101
4.55   L.F. Mollenauer, R.H. Stolen, J.P. Gordon, W.J. Tomlinson: *Opt. Lett.* **8** (1983)
       289
4.56   A. Hasegawa: *Opt. Lett.* **9** (1984) 288
4.57   E.M. Dianov, A. Y. Karasik, P.V. Mamyshev, G.I. Onishchukov, A.M.
       Prokhorov, M.F. Stel'makh, A.A. Fomichev: *JETP Lett.* **40** (1984) 903
4.58   J.N. Sisakyan, A. Shvartsburg: *Kvantov. Elektron.* **11** (1984) 1703
4.59   E.M. Dianov, S.C. Nikonov, W.N. Serkis: Preprint IOFAN No. 13 Moscow
       (1985)
4.60   L.F. Mollenauer, J.P. Gordon, M.N. Islam: *IEEE J. Quant. Electr.* **QE22** (1986)
       157
4.61   K. Tai, A. Tomita, J.L. Jewell, A. Hasegawa: *Appl. Phys. Lett.* **49** (1986) 236
4.62   D. Anderson, M. Lisak: *Phys. Rev.* **A27** (1983) 1393
       C.R. Menyuk: *Opt. Lett.* **12** (1987) 614
       K. Okhuma, Y.H. Ichikawa, Y.Abe: *Opt. Lett.* **12** (1987) 516
4.63   H. Eichhorn: Preprint Friedrich-Schiller-University No. 62–65, Jena (1984)
4.64   A.B. Shvartsburg, I.N. Sissakyan: *Opt. and Quant. Electr.* **16** (1984) 207
4.65   see [1.28]
4.66   A.S. Gouveia-Neto, A.S.L. Gomes, J.R. Taylor: *Opt. Lett.* **12** (1987) 395
4.67   R.H. Stolen, J. Botineau, A. Ashkin: *Opt. Lett.* **7** (1982) 512
4.68   B. Nikolaus, D. Grischkowsky, A.C. Balant: *Opt. Lett.* **8** (1983) 189
4.69   H.G. Winful: *Appl. Phys. Lett.* **47** (1985) 213
4.70   N.J. Halas, D. Grischkowsky: *Appl. Phys. Lett.* **48** (1986) 823
4.71   A. Piskarskas, A. Stabinis, A. Jankauskas: *Kvantov. Elektr.* **12** (1985) 1781
4.72a  A. Piskarskas, D. Podenas, A. Stabinis, A. Umbrasas, A. Varanavichias, A.
       Yankauskas, G. Yonushauskas: in [1.32] p. 142
    b  A. Piskarskas, A. Stabinis, A. Yankauskas: *Sov. Phys. Usp.* **29** (1986) 869
4.73   see [2.3]
4.74   see [2.4]
4.75   T.N.C. Venkatesan, S.L. McCall: *Appl. Phys. Lett.* **30** (1977) 282
4.76   D. Kühlke, W. Rudolph, B. Wilhelmi: *IEEE J. Quant. Electr.* **QE19** (1983) 526
4.77   W. Rudolph, B. Wilhelmi: *Appl. Phys.* **B35** (1984) 37
4.78   D. Kühlke, W. Rudolph: *Opt. Quant. Electr.* **16** (1983) 57
4.80   see [1.7]
4.81   O.E. Martinez, R.L. Fork, J.P. Gordon: *J. Opt. Soc. Am.* **B2** (1985) 753
4.82   Y. Ishida, T. Yajima: *IEEE J. Quant. Electr.* **QE21** (1985) 69
4.83   B. Wilhelmi, W. Rudolph, E. Döpel, W. Dietel: *Opt. Acta* **32** (1985) 1175
4.84   J.J. Fontaine, J.C. Diels, W. Dietel: *IEEE J. Quant. Electr.* **QE19** (1983) 1467
4.85   see [3.12]
4.86   R.S. Miranda, G.R. Jacobowitz, C.H. Brito-Cruz, M.A. Scarparo: *Opt. Lett.* **11**
       (1986) 224

4.87   L. Allen, J.H. Eberly: *Optical Resonance and Two-Level Atoms*, Wiley, New York, London, Sydney, Toronto (1975)
4.88   C.H. Brito Cruz, R.L. Fork, W.H. Knox, C.V. Shank: *Chem. Phys. Lett.* **132** (1986) 34
4.89   A.J. Taylor, D.J. Erskine, C.L. Tang: *Chem. Phys. Lett.* **103** (1984) 430
4.90   M. Mitsunaga, C.L. Tang: *Phys. Rev.* **A35** (1987) 1720
4.91   A.M. Weiner, E.P. Ippen: *Chem. Phys. Lett.* **114** (1985) 456
4.92   W. Rudolph, J.C. Diels: in [1.32] p. 71
4.93a  M.A. Vasileva, J. Vishchakas, V. Gulbinas, V.I. Malyshev, A.V. Masalov, V. Kabelka, V. Syrus: *IEEE J. Quant. Electr.* **19** (1983) 724
   b   V. Petrov, W. Rudolph, B. Wilhelmi: *Opt. Commun.* **64** (1987) 398
   c   V. Petrov, W. Rudolph, B. Wilhelmi: submitted to Kvantov. Elektr.
4.94   M.D. Crisp: *Phys. Rev.* **A8** (1973) 2128
4.95   L.M. Frantz, J.S. Nodvik: *J. Appl. Phys.* **34** (1963) 2346
4.96   W. Dietel, E. Döpel, W. Rudolph, B. Wilhelmi: *Scient. Instr.* **1** (1986) 71
4.97   A. Penzkofer: *Opto-Electronics* **6** (1974) 87
4.98   A. Icsevgi, W.E. Lamb: *Phys. Rev.* **185** (1969) 517
4.99   W. Rudolph: *Opt. Quant. Electr.* **16** (1984) 541
4.100  W. Rudolph, B. Wilhelmi: *Opt. Comm.* **49** (1984) 37
4.101  J.A. Armstrong, E. Courtens: *IEEE J. Quant. Electr.* **QE5** (1969) 249
4.102  R.A. Fischer, W.K. Bischel: *J. Appl. Phys.* **46** (1975) 4921
4.103  P. Heist, V. Petrov, W. Rudolph: subm. to Opt. Comm.
4.104  M. Kaschke, B. Wilhelmi, V.D. Egorov, H.X. Ngyuen, R. Zimmermann: *Appl. Phys. B45* (1987) 71
4.105a R.L. Fork, C.V. Shank, R.Y. Yen: *Appl. Phys. Lett.* **45** (1982) 223
    b  W.H. Knox, M.C. Downer, R.L. Fork, C.V. Shank: *Opt. Lett.* **9** (1984) 552
4.106  W. Dietel, E. Döpel. V. Petrov, G. Marowski, C. Rempel, W. Rudolph, F.P. Schäfer, B. Wilhelmi: *Appl. Phys.* B46 (1988) 183
4.107  Zs. Bor, G. Szabo, F. Raksi: in Abstracts of UPS'87 p. 19, Vilnius (1987)
4.108a D. Strickland, G. Mourou: *Opt. Commun.* **56** (1985) 219
    b  M. Pessot, P. Maine, G. Mourou: *Opt. Comm.* **62** (1987) 419
4.109  Y.J. Chang, C. Veas, J.B. Hopkins: *Appl. Phys. Lett.* **49** (1986) 1758
4.110  A. Hasegawa: *Opt. Lett.* **8** (1983) 650
4.111  E.M. Dianov, A. Y, Karasik, P.V. Mamyshev, G.I. Onishchukev, A.M. Prokhorov, M.F. Stel'makh, A.A. Fomichev: *Pisma Zh. Eksp. Teor. Fiz.* **39** (1984) 564
4.112  R.H. Stolen, C. Lee, R.U. Jain: *J. Opt. Soc. Am.* **B1** (1984) 652
4.113  L.F. Mollenauer, R.H. Stolen, M.N. Islam: *Opt. Lett.* **10** (1985) 229
4.114  E.M. Dianov, A.Y. Karasik, P.V. Mamyshev, A.M. Prokhorov, V.N. Serkin, M.F. Stel'makh, A.A. Fomichev: *Pisma Zh. Eksp. Teor. Fiz.* **41** (1985) 242
4.115  B. Valk, W. Hodel, H.P. Weber: *Opt. Comm.* **54** (1985) 363
4.116  B. Zysset, H.P. Weber: Proceedings CLEO'86, WK 10
4.117  R.H. Stolen, A.M. Johnson: *IEEE J. Quant. Electr.* **QE22** (1986) 2154
4.118  A.M. Johnson, R.H. Stolen, W.M. Simpson: in [1.32] p. 160
4.119  F.M. Mitschke, L.F. Mollenauer: *Opt. Lett.* **11** (1986) 659
4.120  P.M.W. French, A.S.L. Gomes, A.S. Gouveia-Neto, J.R. Taylor: *IEEE J. Quant. Electr.* **QE22** (1986) 2230
4.121  A.M. Weiner, J.P. Heritage, R.H. Stolen: *J. Opt. Soc. Am.* **B5** (1988) 364
4.122  A.S. Gouveia-Neto, A.S.L. Gomes, J.R. Taylor: to appear in *IEEE J. Quant. Electr.* **QE24** (1988) 332
4.123  E.M. Dianov, A.B. Grudinin, D.V. Khaidarov, D.B. Korobkin: in Abstracts of UPS'87 p. 87, Vilnius (1987)
4.124  B. Zysset, P. Beaud, W. Hodel: *Appl. Phys. Lett.* **50** (1987) 1027
4.125  V.A. Vysloukh, V.N. Serkin: *Pisma Zh. Eksp. Teor.·Fiz.* **38** (1983) 170
4.126  A. Hasegawa: *Appl. Opt.* **23** (1984) 3302

| 4.127 | J.P. Gordon: *Opt. Lett.* **11** (1986) 662 |
|---|---|
| 4.128 | W. Hodel, H.P. Weber: submitted to Opt. Comm. |
| 4.129 | J. Herrmann: to be published |
| 4.130 | D. Schadt, B. Jaskorzynska, U. Österberg: *J. Opt. Soc. Am.* **B3** (1986) 1257 |
| 4.131 | K.J. Blow, N.J. Doran, D. Wood: submitted to *Phys. Rev. Lett.* |
| 4.132 | L.J. Hua, L.Y. Lin, J.J. Lin: *Opt. Quant. Electr.* **17** (1985) 187 |
| 4.133 | R.H. Stolen, E.P. Ippen: *Appl. Phys. Lett.* **22** (1973) 276 |
| 4.134 | R.G. Smith: *Appl. Opt.* **11** (1972) 2483 |
| 4.135 | L.G. Cohen, C. Lin: *IEEE J. Quant. Electr.* **QE14** (1978) 855 |
| 4.136 | C. Lin, R.H. Stolen: *Appl. Phys. Lett.* **29** (1976) 428 |
| 4.137 | see [2.4] |

| 5.1 | E.B. Treacy: *Phys. Lett.* **28A** (1986) 34 |
|---|---|
| 5.2 | E.P. Ippen, C.V. Shank: *Appl. Phys. Lett.* **27** (1975) 488 |
| 5.3 | see [1.6] |
| 5.4 | see [4.84] |
| 5.5 | J.C. Diels, W. Dietel, J.J. Fontaine, W. Rudolph, B. Wilhelmi: *J. Opt. Soc. Am.* **B2** (1985) 680 |
| 5.6 | J.A. Valdmanis, R.L. Fork: *IEEE J. Quant. Electr.* **QE22** (1986) 112 |
| 5.7 | see [1.12] |
| 5.8 | D. Külke, T. Bonkhofer, U. Herpers, D. von der Linde: in [1.32] p. 17 |
| 5.9 | see [3.17] |
| 5.10 | W. Dietel, W. Rudolph, B. Wilhelmi, J.C. Diels, J.J. Fontaine: *Isv. Akad. Nauk SSSR* **48** (1984) 480 |
| 5.11 | see [4.77] |
| 5.12 | see [4.81] |
| 5.13a | F. Salin, P. Grangier, G. Roger, A. Brun: *Phys. Rev. Lett.* **56** (1986) 1132 |
| b | H. Avramopoulos, P.M.W. French, J.A.R. Williams, G.H.C. New, J.R. Taylor: *IEEE J. Quant. Electr.* **QE24** (1988) |
| 5.14 | L.F. Mollenauer, R.H. Stolen: *Opt. Lett.* **9** (1984) 13 |
| 5.15 | F.M. Mitschke, L.F. Molenauer: *IEEE J. Quant. Electr.* **QE22** (1986) 2242 |
| 5.16 | G.A. Mourou, T. Sizer II: *Opt. Comm.* **41** (1982) 47 |
| 5.17 | J.M. Halbout, C.L. Tang: *IEEE J. Quant. Electr.* **QE19** (1983) 487 |
| 5.18 | H. Vanherzeele, J.C. Diels, R. Torti: *Opt. Lett.* **9** (1984) 549 |
| 5.19 | H. Vanherzeele, R. Torti, J.C. Diels: *Appl. Opt.* **23** (1984) 4128 |
| 5.20 | J. Dobler, H.H. Schulz, W. Zinth: *Opt. Comm.* **57** (1986) 407 |
| 5.21 | G.R. Jacobowitz, C.H. Brito Cruz, M.A. Scarparo: *Opt. Comm.* **57** (1986) 133 |
| 5.22 | M.D. Dawson, T.F. Boggess, D.W. Garvey, A. Smirl: *IEEE J. Quant. Electr.* **QE22** (1986) 2195 |
| 5.23 | J.C. Diels, N.Jamasbi, L. Sarger: in [1.32] p. 2 |
| 5.24 | P.M.W. French, J.R. Taylor: *Opt. Lett.* **11** (1986) 297 |
| 5.25 | P.M.W. French, J.A.R. Williams, J.R. Taylor: *Revue Phys. Appl.* **22** (1987) 1651 |
| 5.26 | J. Chesnoy, L. Fini: in [1.32] p. 14 |
| 5.27 | J.S. Ruddock, D.J. Bradley: *Appl. Phys. Lett.* **29** (1976) 296 |
| 5.28 | see [1.7] |
| 5.29 | W. Dietel: *Opt. Comm.* **43** (1982) 69 |
| 5.30 | W. Dietel: E. Döpel, D. Kühlke, B. Wilhelmi: *Opt. Comm.* **43** (1982) 433 |
| 5.31a | V. Petrov: PhD Thesis, Friedrich-Schiller-University Jena, (1987) |
| b | A. Finch, G. Chen, W. Sleat, W. Sibett: J. Mod. Opt. 35 (1988) 345 |
| 5.32 | see [3.11] |
| 5.33 | see [3.7] |
| 5.34 | O.E. Martinez: *Opt. Comm.* **59** (1986) 229 |
| 5.35 | J.A. Fleck: *J. Appl. Phys.* **39** (1968) 3318 |
| 5.36 | G.H.C. New: *IEEE J. Quant. Electr.* **QE10** (1974) 115 |
| 5.37 | G.H.C. New, D.H. Rea: *J. Appl. Phys.* **47** (1976) 3107 |

128    W. RUDOLPH and B. WILHELMI

5.38   B.K. Garside, T.K. Lin. *Opt. Comm.* **12** (1974) 240
5.39   H.A. Haus: *IEEE J. Quant. Electr.* **QE11** (1975) 736
5.40   J. Herrmann, F. Weidner: *Appl. Phys.* **B27** (1982) 105
5.41   J. Herrmann, F. Weidner, B. Wilhelmi: *Appl. Phys.* **B26** (1981) 197
5.42   D. Kühlke, W. Rudolph, B. Wilhelmi: *Appl. Phys. Lett.* **42** (1983) 325
5.43   M.S. Stix, E.P. Ippen: *IEEE J. Quant. Electr.* **QE19** (1983) 520
5.44   M. Yoshizawa, T. Kobayashi: *IEEE J. Quant. Electr.* **QE20** (1984) 797
5.45   D. Kühlke, W. Rudolph, B. Wilhelmi: *IEEE J. Quant. Electr.* **QE19** (1983) 526
5.46   see [4.78]
5.47   W. Rudolph: PhD Thesis, Friedrich-Schiller-University Jena (1984)
5.48   see [4.86]
5.49   O.E. Martinez, R.L. Fork, J.P. Gordon: *Opt. Lett.* **9** (1984) 156
5.50   H.A. Haus, Y. Silberberg: *IEEE J. Quant. Electr.* **QE22** (1986) 325
5.51   M.S. Stix: *Opt. Lett.* **10** (1985) 279
5.52   V. Petrov, W. Rudolph, B. Wilhelmi: *Opt. Quant. Electr.* **19** (1987) 377
5.53   V. Petrov, W. Rudolph, B. Wilhelmi: *Revue Phys. Appl.* **22** (1987) 1639
5.54   D.N. Dempster, T. Morrow, R. Rankin, G.F. Thompson: *J. Chem. Soc. Faraday Trans. II* **68** (1972) 1479
5.55   S. Rentsch, E. Döpel, V. Petrov: *Appl. Phys.* **B46** (1987) 357
5.56   V. Petrov, W. Rudolph, U. Stamm, B. Withelmi: subm. to Phys. Rev. A
5.57   F.M. Mitschke, L.F. Mollenauer: *Opt. Lett.* **11** (1986) 659
5.58   H.A. Haus, M.N. Islam: *IEEE J. Quant. Electr.* **QE21** (1985) 1172
5.59   F.If, P.L. Christiansen, J.N. Elgin, J.G. Gibbon, O. Skovgaard: *Opt. Comm.* **57** (1986) 350
5.60   K.J. Blow, D. Wood: *IEEE J. Quant. Electr.* **QE22** (1986) 1109
5.61   P. Berg, F.If, P.L. Christiansen, O. Skovgaard: *Phys. Rev.* **A35** (1987) 4167
5.62   see [4.10]
5.63   see [4.53]
5.64   L.F. Mollenauer, N.D. Nieira, L. Szeto: *Opt. Lett.* **7** (1982) 414
5.65   L.G. Cohen, A.D. Pearson: *Proc. Soc. Photo-Opt. Instrum. Eng.* **425** (1983) 28
5.66   Z.A. Yasa, O. Teschke: *Opt. Comm.* **15** (1975) 169
5.67   C.P. Ausschnitt, R.K. Jain, J.P. Heritage: *IEEE J. Quant. Electr.* **QE15** (1979) 912
5.68   J.M. Catherall, G.H.C. New, P.M. Radmore: *Opt. Lett.* **7** (1984) 49
5.69   J. Hermann, U. Motschmann: *Appl. Phys.* **B27** (1982) 27
5.70a  D. Schubert, U. Stamm, B. Wilhelmi: *Opt. Quant. Electr.* **17** (1985) 337
   b   U. Stamm: *Appl. Phys.* **B45** (1988) 101
   c   U. Stamm, F. Weidner, B. Wilhelmi: *Opt. Comm.* **63** (1987) 179
5.71   J.M. Catherall, G.H.C. New: *IEEE J. Quant. Electr.* **QE22** (1986) 1593
5.72   L.F. Mollenauer: Proc. of the Workshop on Structure, Coherence and Chaos in Dynamical Systems, Lungby (1986)
5.73   R.H. Stolen: *Fiber and Integrated Optics* **3** (1980) 21
5.74   A. Hasegawa: *Opt. Lett.* **8** (1983) 650
5.75   L.F. Mollenauer, R.H. Stolen, M.N. Islam: *IEEE J. Quant. Electr.* **QE22** (1986) 157
5.76   B. Zysset, P. Beaud, W. Hodel, H.P. Weber: in [1.23] p. 54
5.77   J.D. Kafka, T. Baer: *Opt. Lett.* **12** (1987) 181
5.78   A.S. Gouveia-Neto, A.S.L. Gomes, J.R. Taylor: *IEEE J. Quant. Electr.* **QE23** (1987) 1193

6.1    see [1.13b]
6.2a   see [1.1]
   b   K.L. Sala, G.A. Kenney-Walace, G.E. Hall: *IEEE J. Quant Electr.* **QE16** (1980) 990
6.3    A.M. Weiner: *IEEE J. Quant. Electr.* **QE19** (1983) 1276
6.4    J.C. Diels, E.W. Van Stryland, D. Gold: in Picosecond Phenomena I, p. 117, Springer, New York (1981)
6.5    J.C. Diels, J.J. Fontaine, I.C. McMichael, F. Simoni: *Appl. Opt.* **24** (1985) 1270

6.6     see [5.23]
6.7a    J.E. Rothenberg, D. Grischkowsky: *J. Opt. Soc. Am.* **B2** (1985) 626
   b    J.E. Rothenberg: *IEEE J. Quant. Electr.* **QE22** (1986) 174
6.8     N. Bloembergen, E.M. Purcell, R.V. Pound: *Phys. Rev.* **73** (1948) 679
6.9     B.A. Jacobsohn, R.K. Wangsness: *Phys. Rev.* **73** (1948) 942
6.10    M.D. Crisp: *Phys. Rev.* **A1** (1970) 1604
6.11    J.E. Rothenberg, D. Grischkowsky, A.C. Balant: *Phys. Rev. Lett.* **53** (1984) 552
6.12a   J.E. Rothenberg, D. Grischkowsky: *J. Opt. Soc. Am.* **B3** (1986) 1235
    b   G. Werner: Diploma Thesis, Friedrich-Schiller-University Jena (1987)
6.13    see [3.13]
6.14    see [3.11]
6.15    J.C. Diels, W. Dietel, J.J. Fontaine, I.C. McMichael, W. Rudolph, F. Simoni, B. Wilhelmi: Proceedings of IQEC (1984)
6.16    J.C. Diels, I.C. McMichael: *J. Opt. Soc. Am.* **B3** (1986) 535
6.17    A.M. Weiner, E.P. Ippen: *Opt. Lett.* **9** (1984) 53
6.18    J.G. Fujimoto, E.P. Ippen: *Opt. Lett.* **8** (1983) 446
6.19    W.Z. Lin, J.G. Fujimoto, E.P. Ippen: *Appl. Phys. Lett.* **50** (1987) 124
6.20    J.P. Heritage, R.N. Thurston, W.J. Tomlinson, A.M. Weiner, R.H. Stolen: *Appl. Phys. Lett.* **47** (1985) 87
6.21    R.N. Thurston, J.P. Heritage, A.M. Weiner, W.J. Tomlinson: *IEEE J. Quant. Electr.* **QE22** (1986) 682
6.22    A.M. Johnson, W.M. Simpson: *IEEE J. Quant. Electr.* **QE22** (1986) 133

# INDEX